S0-BZZ-758

Fingerprint Detection
With LASERS

Fingerprint Detection With LASERS

E. ROLAND MENZEL

Department of Physics
Texas Tech University
Lubbock, Texas

MARCEL DEKKER, INC.   New York and Basel

Library of Congress Cataloging in Publication Data

Menzel, E. Roland [date]
Fingerprint detection with lasers.

Includes bibliographical references and index.
1. Fingerprints--Laser use in. I. Title
HV6074.M44   363.2'58   80-14159
ISBN 0-8247-6974-0

MARCEL DEKKER, INC.
270 Madison Avenue, New York, New York 10016

Current printing (last digit):
10 9 8 7 6 5 4 3

PRINTED IN THE UNITED STATES OF AMERICA

To Ingeborg and Theresa

PREFACE

The application of lasers to detection of latent fingerprints, initiated in 1976, has already become instrumental in a number of criminal investigations and can be anticipated to develop into a highly valuable forensic tool. Indeed, several law enforcement agencies now use lasers for latent fingerprint detection, and others are in the process of acquiring lasers for this purpose. Court cases involving laser-detected latent prints are presently pending. Two manufacturers in the United States now offer laser systems specifically designed for fingerprint detection.

This text aims to provide forensic researchers and investigators with a guide to both established and potential uses of lasers and optical spectroscopy in forensic identification. Forensic analysts will benefit from having the available information on the current state of laser detection of latent prints collected in one text, rather than having to search the scattered, and often terse, research literature. A text at this stage is timely, since lasers and spectroscopic instrumentation are expensive, so that some forensic laboratories may be reluctant to consider acquisition of this machinery unless cognizant in detail of its already demonstrated as well as potential utility.

Optical spectroscopic techniques are valuable in optimizing the sensitivity of laser detection of latent prints. In discussing optical spectra, the principles of operation of lasers, etc., it is necessary to deal with quantum mechanical aspects in order to gain fundamental insight. Chapter 1 approaches this in a highly simplified form. The chapter is designed to give the reader, who in the majority of cases will likely have little formal training in the physical sciences, some appreciation of the nature of atomic and molecular spectra. Luminescence is particularly emphasized, since it is the crux of laser detection of latent prints. A description of the principles of laser operation is also provided. For readers who want a more thorough treatment of the subject, supplementary references are given at the end of the chapter. A major aim of Chapter 1 is to acquaint latent print examiners with the basic physical principles underlying laser detection of latent prints, since questions about these principles can be anticipated to arise during court testimony.

While there are a number of ways in which lasers have already been used in forensic work, a great deal of research has yet to be undertaken if lasers are to be exploited to their full potential in forensic analysis. Such research will often require measurement of absorption, luminescence, and luminescence excitation spectra. Chapter 2 describes the general features of instrumentation for measurement of such spectra. Since an extensive treatment of a large variety of spectroscopic techniques is beyond the scope of the text, the description of instrumentation is restricted to a flexible and cost-effective modular spectroscopic system whose components can be rearranged for a variety of tasks. The system, while specifically geared for use in latent fingerprint luminescence studies, is useful for optical spectroscopy on a much wider scope. Readers who wish to approach the utilization of lasers in forensic work pragmatically and who do not engage in research should be able to ignore parts of Chapters 1 and 2. Some readers may find it helpful to glance at Chapters 3 and 4 prior to reading Chapters 1 and 2.

The specific application of lasers to the development of laser fingerprints is discussed in Chapter 3. Latent print detection by inherent fingerprint luminescence is discussed and a variety of latent print treatments which lead to laser-detected luminescence are then described. Case applications which provide a measure of the power of several classes of procedures are presented. Laser detection via inherent luminescence requires *no* physical or chemical treatment of the exhibit under scrutiny. Latent print treatments yielding luminescence use procedures analogous to or identical with conventional methods and, at the same time, take advantage of the great detection sensitivity lasers can provide.

The use of optical spectroscopy in latent fingerprint detection by laser is described in Chapter 4. Measurements discussed in this chapter serve two purposes: optimization of detection sensitivity and investigation of the nature of luminescers in fingerprint residue. Filters needed to optimize detectability are explicitly treated, since correct filter selection is often critical to laser detection of latent prints. Finally, Chapter 4 briefly considers a number of potential uses of lasers and optical spectroscopy in forensic work.

Any user should be familiar with the operation and maintenance of argon-ion lasers. This subject is briefly addressed in Chapter 5. It is intended to supplement instruction manuals for lasers, and only considers the most salient features of laser operation and maintenance.

The research which forms the basis for this text was undertaken at the Xerox Research Centre of Canada. I am greatly indebted to Michael L. Hair (XRCC) for a critical reading of the manuscript, a number of very helpful suggestions, and for interceding on my behalf with Marcel Dekker, Inc.

In return for royalties, Xerox Research Centre of Canada has underwritten the production of the color photographs of the text.

E. Roland Menzel

CONTENTS

1 LIGHT, SPECTRA, AND LASERS

In 1976, development of a method for detection of latent finger-prints, using a continuous wave argon-ion laser, was initiated at the Xerox Research Centre of Canada. In the early stages of this development, the detection of latent prints proceeded primarily via their inherent luminescence [1]. In a nutshell, detection amounted to illumination of exhibits by Ar-laser light and observation through suitable filters of luminescence from latent finger-prints. These prints, once observed via their luminescence, were photographed, again through appropriate filters. Even at this stage the method sparked sufficient interest by several law enforcement agencies to cause them to acquire Ar-lasers for fingerprint work. Since then, a number of procedures for laser detection of latent prints which are not amenable to detection by inherent lumin-escence have been developed [1-5]. These procedures entail treatments of exhibits which yield laser-detected luminescence. Often such treatments are analogous to conventional ones. At the same time, they take advantage of the greatly enhanced detection sensitivity lasers and luminescence photography can provide. For instance, luminescent dusting powders can be employed. If a dusted print is not observable in room light, then laser examina-tion may bring out the elusive ridge detail. Similarly, chemical reagents analogous to ninhydrin can be used.

The luminescent materials in fingerprint residue or the treat-ment compounds which give rise to luminescence are organic molecules in most instances. We will, therefore, largely confine our attention in this chapter to organic molecules. The organic mole-

1

cules of interest to us are usually comprised of at least tens of atoms, such as carbon, hydrogen, nitrogen, and oxygen, which are linked together by chemical bonds. The detection of latent fingerprints by laser amounts in essence to incidence of light on such molecules, absorption of this light by them, and subsequent emission of light different in color from the incident light by them. Molecules require illumination with light of well-defined colors if luminescence is to be obtained. The luminescence also is of a specific color. For luminescence to occur, light must first be absorbed by these molecules. The determination of the light colors which are best suited for absorption is done by absorption or luminescence excitation spectroscopy. The determination of the colors of luminescence is performed by measurement of luminescence spectra. Absorption, excitation, and luminescence spectra are necessary, at least in the research stage, if laser detection of latent prints is to be exploited to its maximum sensitivity. For example, the fingerprint luminescence color, as compared to the color of background luminescence, which often occurs, dictates the kind of filter(s) necessary for best photographic contrast. As will be seen in Chapter 4, spectroscopic techniques can also permit one to locate latent prints which are too weak to be seen by eye under the laser. Once located spectroscopically, such prints can be photographed. This chapter deals with the nature of light, features of light absorption and emission by molecules, and, finally, the principles of operation of lasers. The chapter is partly designed to provide background which will allow latent print examiners to field questions one might expect to arise during court testimony about the physical principles involved in laser detection of latent prints.

In a fundamental sense, insight into the nature of atomic and molecular spectra, into the principles of operation of lasers, indeed into the very nature of the light absorption and emission processes, requires a "quantum mechanical" approach. Unfortunately, quantum theory, by which the interpretation of atomic and molecular spectra is made, entails a great deal of mathematical complexity and conceptual difficulty. This text assumes that most readers will not have much formal training in the physical sciences. Quantum mechanical aspects of spectra are therefore described in a highly simplified form. The purpose of this description is largely to provide a "flavor" of the nature of atomic and molecular spectra (since laser detection of latent prints is in essence luminescence spectroscopy) and to delineate some of the terminology which recurs in the text and, for that matter, throughout the spectroscopic literature. Readers who want an in-depth understanding of atomic and molecular spectra are referred to the supplementary references at the end of the chapter.

1.1 LIGHT

One usually thinks of light as wavelike in nature. The wavelike nature of light is demonstrated, for instance, by the phenomenon of diffraction (light bending, as in the passage of light from air into water). Waves can be of several types. Sound waves, for example, are characterized by back-and-forth vibration of molecules in the direction of propagation of the sound wave. Molecules can also vibrate in a direction perpendicular to the direction of propagation of waves, as in water waves. Finally, there are electromagnetic waves, such as radio waves. Let us briefly consider the nature of electromagnetic waves in some detail. Electric charges have an electric "field" around them by which opposite charges are attracted and like charges are repelled, just as magnets have a magnetic field by which opposite magnetic poles are attracted and like magnetic poles are repelled. Movement of electric charge causes a magnetic field to be set up (which is how electromagnets operate). Electric and magnetic fields have magnitude (strength) and direction. Suppose, for example, we have two opposite charges vibrating periodically against each other. This causes the electric field between them to oscillate in strength and direction. It also produces an oscillating magnetic field which has a direction perpendicular to the electric field at every point in space. The thus produced electromagnetic oscillation propagates as a wave through space with a constant velocity of 3×10^8 m/sec. We will use exponential notation in this text: $3 \times 10^8 = 300,000,000$ (eight zeros). Radio waves and microwaves are examples of electromagnetic waves. In addition to speed, waves are characterized by wavelength and frequency. If one were to measure the strength and direction of the electric field at a single time at various points along the direction of propagation of an electromagnetic wave, one would find a pattern of the electric field as shown in Figure 1.1, where field direction is denoted by arrow direction and field strength is denoted by arrow length. The distance between crests of this sinusoidal wave pattern is the wavelength. If, instead, one were to sample the electric field strength and direction at a single point in space, but at various times, one would again obtain a pattern as in Figure 1.1, but with distance scale replaced by time scale. The interval between crests would now denote the frequency with which crests occur, i.e., how many times per second the wave crests at a given point. The speed, c, the wavelength, λ, and the frequency, ν, are connected by the relation

$$c = \lambda\nu \qquad\qquad\qquad (1.1)$$

Since the speed of the wave is a known constant, it suffices to determine either the wavelength or the frequency to specify the main characteristics of a wave. Light waves are electromagnetic waves, identical in nature to radio waves or microwaves, differing from them only in wavelength and frequency. The speed of light propagation, c, is again 3×10^8 m/sec. The most important fea-

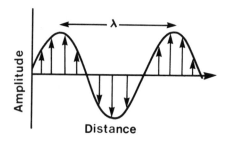

Figure 1.1 Schematic representation of an electromagnetic wave. Vertical arrows denote electric field strength by their length and electric field direction by their direction.

ture of a light wave, namely its color, is customarily specified in units of wavelength rather than frequency. The usual units of light wavelength are microns (or micrometers), μm, angstroms, Å, or nanometers, nm. Values of these units in terms of meters are

$$1 \, \mu m = 10^{-6} \, m$$
$$1 \, \overset{\circ}{A} \; = 10^{-10} \, m$$
$$1 \, nm = 10^{-9} \, m$$

For the description of a number of physical phenomena it is more appropriate to consider light to be particle-like rather than wavelike. This is not just a matter of conceptual convenience; it is a consequence of the quantum mechanical nature of light. Indeed, wave-particle duality is not confined to light. Electrons, for instance, which one is accustomed to think of as particles, display diffraction effects which are wave phenomena. For purposes of description of the nature of atomic and molecular spectra, as well as the operation of lasers, the particle-like nature of light is highly pertinent. Light particles, or photons (sometimes called quanta), have energy, E, given by

$$E = h\nu \tag{1.2}$$

where h is Planck's constant, of value 6.63×10^{-34} in units of joules $\times$ second, and where 1 joule/second = 1 watt. Negative exponents have the meaning given by the example $10^{-5} = 1/100,000$ (five zeros in the denominator). In units of watts, h = 6.63×10^{-34} watt $\times$ second^2. The quantity ν is none other than the light frequency of equation 1.1. In terms of wavelength we thus have

$$E = \frac{hc}{\lambda} \tag{1.3}$$

This means that the characteristic feature of light, namely its color, can be specified either by wavelength or by photon energy. Photon energies are usually given in units of electron volts (eV) or wave numbers (cm^{-1}). To obtain the energy in wave numbers, one

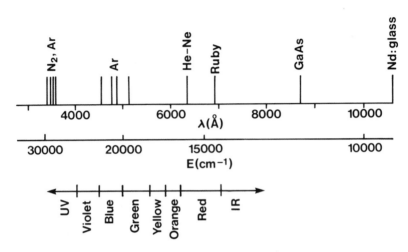

Figure 1.2 Wavelength and energy range of the visible spectrum and wavelengths of some lasers.

takes the light wavelength, λ, in units of centimeters (cm) and computes $1/\lambda$. One electron volt equals approximately 8000 cm^{-1}. In terms of photons, a monochromatic (single color) light beam of wavelength λ is now nothing more than a group of photons, all of which have the same energy, hc/λ, with beam intensity corresponding to number of photons. For example, a 10 W beam of the blue-green light from an Ar-ion laser illuminating a surface for 1 sec causes the surface to be struck by some 3×10^{19} photons, each of which has an energy of approximately 20,000 cm^{-1}. Unlike ordinary light sources, which radiate a broad range of colors and emit in all directions, lasers emit a well-collimated (parallel rays) light beam which is either monochromatic or consists of a few "lines," i.e., a beam consisting of a few very sharply defined colors. Figure 1.2 shows the wavelength range of the visible spectrum and light wavelengths of several lasers.

1.2 ATOMS, MOLECULES, AND SPECTRA

For our purposes, it suffices to consider an atom as comprised of a nucleus containing a number of elementary particles, most notably protons (positively charged), about which electrons (negatively charged) orbit in a manner somewhat analogous to planets orbiting a sun. In the classical mechanics framework, i.e. a conceptual framework which deals with physical phenomena on a macroscopic scale, the energy (potential and kinetic) of an orbiting planet is specified by the mass of the planet (and, of course, the mass of the sun), its velocity, and the radius of its orbit (if circular). This energy can vary "continuously," i.e., arbitrarily small energy changes are possible. Atoms and molecules, however, are quantum mechanical systems. Orbits in which electrons can be found (atomic and molecular orbitals) can only take on "discontinuously" varying energy values, i.e., *discrete* energies. An electron can be promoted from a given orbital to a higher energy orbital by the *absorption* of *one* photon. The photon is gobbled up by the electron which thus acquires the necessary energy to jump to the higher orbital. Conversely, an electron can jump from a higher orbital to a lower one and release the excess energy in the form of *one emitted* photon. The photon energy, or light color, involved in the acts of absorption or emission corresponds to the difference in energies of the involved orbitals, E, and the absorbed or emitted photon has wavelength hc/E. The wavelengths of light which can be absorbed or emitted by an atomic or molecular system, i.e. the absorption or emission spectra, probe the "energy level" structure (or "states") of the system. These states are characteristic of the species under study, just as fingerprints are characteristic of individuals. In atomic spectroscopy, these wavelengths are very sharply defined and "line" spectra are measured. The measured line widths give the precision with which orbital energy differences are determined. Line widths in atomic spectroscopy are typically on the order of 10^{-1} to 1 Å. In molecules, electrons are not confined to orbit a single nucleus, but can spread out over several nuclei. Consequently, a greater number of orbitals are available. These are often closely spaced and include orbitals which differ only by discrete vibrational and rotational states of the molecule. Moreover, because of the greater spacial extent of molecular orbitals, compared to atomic orbitals, electrons in molecules are susceptible to a variety of perturbations which significantly affect orbital energies. As a result, molecular spectra are very often much broader than atomic spectra, with line widths of hundreds, sometimes even thousands, of angstroms.

In absence of external influences, electrons in molecules reside in orbitals of the lowest possible energy. This state of the molecule is called the *ground state*. Promotion of an electron to a higher energy orbital, by absorption of light, electric discharge, etc., results in an *excited state* of the atom or molecule. Atoms and molecules in excited states tend to quickly return to the ground state (principle of universal laziness) by emission of light or other means of releasing energy. Emission of light (luminescence) during this return to the ground state can take on the form of *fluorescence* or *phosphorescence*. For the moment it suffices to note that molecular fluorescence takes place very quickly, generally in a time less than 10^{-6} sec, after absorption of light has generated an excited state. Phosphorescence, on the other hand, occurs in molecules generally 1 to 10^{-5} sec after absorption, and phosphorescence occurs in a given molecule at a lower energy (longer wavelength) than fluorescence. Molecules seldom display both pronounced fluorescence and phosphorescence simultaneously, however. While molecules in solution, particularly at low temperatures, frequently show intense luminescence, molecular solids, such as powders, very rarely show observable phosphorescence at room temperature and usually fluoresce only weakly.

In contrast to atoms, molecules slightly rearrange their structure once absorption has promoted an electron to a higher orbital. This rearrangement is due to change in electrostatic forces between electrons and nuclei of the molecule arising from the change in electron distribution in the molecule caused by the absorption event. The rearrangement takes place before light emission occurs and causes a reduction in energy of the excited state. In terms of absorption and fluorescence, this means that absorption into a given state occurs at higher energy (shorter wavelength) than fluorescence from this state. This wavelength difference, called the Stokes shift, is schematically shown in Figure 1.3.

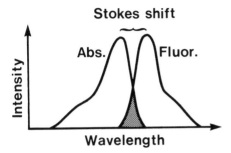

Figure 1.3 Schematic diagram of Stokes shift in molecular spectra.

1.3 MOLECULAR SPECTRA: A CLOSER LOOK

In this section, some features of optical spectra of organic molecules are briefly outlined. Some of the terminology frequently encountered in the spectroscopic literature is also defined.

Let us first consider the arrangements of electrons in molecules in their ground state. Molecular orbitals are filled up by electrons in pairs, with the lowest energy orbitals filled first. Electrons occupy the lowest possible orbitals as long as the molecule is in the ground state. Since the organic molecules of interest to us have an even number of electrons, the highest occupied orbital of a molecule in the ground state must contain two electrons. Optical absorption promotes an electron from this orbital to an unoccupied orbital. For many purposes, a molecule can thus be treated as if it only had two electrons. In the ground state, these two electrons are in the same orbital. In an excited state, one electron remains in this orbital and the other is located in a higher orbital which it occupies alone.

The motion of an electron in its orbital produces a magnetic field. The electron itself has properties of an "elemental" magnet, called the electron *spin.* The total energy of an electron in an orbital depends on the spin, since the magnetic field due to the electron's orbital motion interacts with the electron's spin, just as electromagnets interact with magnetic materials. The total energy

of the electron depends on the direction of the spin with respect
to the direction of the orbital. In nature, the same orbital never
contains more than two electrons. Moreover, if two electrons are
in the same orbital, their spins must be in opposite directions
(antiparallel). This is the Pauli exclusion principle. A state described by two electrons with antiparallel spins is called a *singlet* state.
Since the two electrons in the top-filled orbital of a molecule in
the ground state must have antiparallel spins to satisfy the exclusion principle, the ground state is a singlet state. When light is absorbed by a molecule, one of these two electrons is promoted to
a new orbital. The two electrons, namely the one left behind and
the one promoted to the new orbital, can still have antiparallel
spins, in which case the excited state is a singlet state, or they can
have parallel spins, in which case the state is a *triplet* state.

Optical transitions (light absorption or emission) do not occur
between all states a molecule can possibly assume. Optical transitions are "allowed," i.e. are strong transitions, only as long as no
spin flip (one electron changing its spin direction so that spins
which were antiparallel become parallel) occurs during the transition. Optical transitions in which a spin flip occurs are "forbidden,"
i.e., are weak or do not occur at all. An allowed transition whereby an electron jumps from a higher orbital to the ground state
orbital with the emission of light is called fluorescence. Radiationless transitions, in which the jump of an electron from one orbital
to another is not accompanied by the emission of light occur in
molecules. Such transitions can allow electron jumps such that a
molecule goes from an excited singlet state to a triplet state. From
a triplet state, return to the ground state by light emission is possible. Such a transition is called phosphorescence. Since a spin flip
occurs during phosphorescence, this transition is not very probable
and the return to the ground state occurs on a time scale much
longer than fluorescence.

Each molecular excited singlet state has a triplet counterpart,
usually at lower energy. The ground state (singlet) does not have
such a counterpart because of the exclusion principle. This situation is depicted in Figure 1.4. The absorption spectrum probes
transitions from the ground state to the various excited singlet

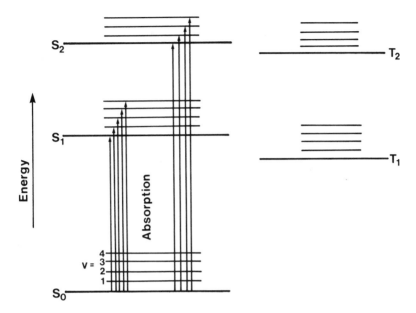

Figure 1.4 Molecular energy level diagram and states accessible to light absorption. S_0 = ground state; S_1 = first excited singlet state; T_1 = lowest triplet state (see text).

states, labeled by subscripts 0, 1, 2 in increasing energy, but not to triplet states. As mentioned earlier, molecules can also change their vibrational configurations, i.e., the different ways in which nuclei can vibrate with respect to each other. Such vibrational states are labeled as v = 1 to 4 in Figure 1.4. Radiationless transitions between excited singlet states including vibrational sublevels are generally extremely rapid, much faster than fluorescence. Consequently, absorption into an excited singlet state is followed by radiationless relaxation to the lowest level of the first excited singlet state vibrational manifold as shown in Figure 1.5 by waved arrows. From this level, fluorescence to the ground state vibrational manifold can occur (straight downward arrows). Alternatively, radiationless relaxation to the triplet manifold, called intersystem crossing, can occur. Again, radiationless relaxation within levels of the triplet manifold is very rapid, so that phosphorescence occurs

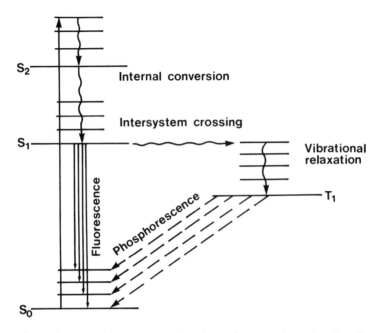

Figure 1.5 Radiative and radiationless relaxation in molecules. Waved arrows denote radiationless relaxation paths. The straight upward arrow denotes light absorption.

from the lowest triplet level (broken arrows in Figure 1.5). Very often, the vibrational energy level structure of the ground state is very similar to that of the first excited singlet state. This similarity gives rise to mirror symmetry between absorption and fluorescence, as shown in Figure 1.6.

One should note that intersystem crossing can dominate over fluorescence. Also, radiationless relaxation from the lowest triplet state to the ground state can be fast, dominating over phosphorescence. Many substances, therefore, show neither fluorescence nor phosphorescence.

When an absorption spectrum is taken, light intensity I_0 illuminates the sample and the light intensity I which is transmitted through the sample (at a given wavelength, that is) is measured.

The optical density (OD) of the sample at this wavelength is given by

$$OD = \log_{10} \frac{I_0}{I} \qquad (1.4)$$

where $\log_{10}$ means base 10 logarithm. Commercial absorption spectrophotometers directly display optical density on the chart recorder which records the absorption spectrum. Knowing the concentration, c, of absorbing molecules in the sample and the optical light path, l, i.e. the sample thickness, the "transition strength" at a given wavelength can be obtained from the optical density at that wavelength. This transition strength is called the molar extinction coefficient, ξ, and is given by

$$\xi = \frac{OD}{cl} \qquad (1.5)$$

where the concentration is given in molar units and the optical path in centimeters. Optical density has no units. Molecules have a molecular weight which specifies the mass of the molecular species of interest. Molecular weights (often denoted as MW on containers of chemicals) are given as numbers without units. For example, the molecular weight of methanol is 32. If M grams of a compound of molecular weight M are dissolved in a liter of solvent, then the concentration of the solute is 1 molar. If m grams of the

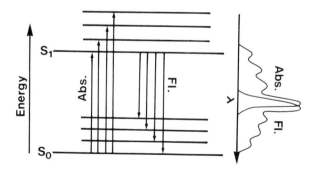

Figure 1.6 Energy level scheme showing origin of mirror symmetry between absorption and fluorescence.

compound are dissolved in X liters of solvent, the concentration is
mX/M in molar units.

Highly colored pigments and dyes tend to have high extinction coefficients at their wavelength of maximum absorption. For
example, laser dyes such as coumarin 6 and rhodamine 6G, which
are discussed in Chapter 3, have extinction coefficients on the order of 10^5. Often dyes with such high extinction coefficients are
strongly fluorescent and may be useful for latent fingerprint treatment followed by laser examination. Indeed, in many cases the
extinction coefficient of a compound provides a guide to the
strength with which one can expect it to fluoresce. Optical density
determination is required for measurements of luminescence quantum yields (the probability that an absorbed photon will give rise
to a luminescent photon). This is discussed briefly in Chapter 2.

1.4 PRINCIPLES OF LASER OPERATION

In conventional incandescent lamps, energy is pumped into atoms
of the lamp filament electrically. These atoms are thus raised to
excited states from which they drop back to the ground state
spontaneously, emitting randomly directed photons. The atoms in
the filament radiate independently both in terms of direction and
time. The radiated light is thus *incoherent*. Fluorescence and phosphorescence are generally incoherent, spontaneous emission. Light
emission from atoms or molecules in excited states cannot only
occur spontaneously, however, but can be triggered as well. Suppose we have an atom in an excited state which differs in energy
from the ground state by a value E. If a photon of energy $h\nu = E$
impinges on this excited atom, it can trigger the relaxation of this
atom back to the ground state. The atom thus triggered also re-
leases a photon of energy $h\nu = E$. Moreover, the released photon
has the same phase (wave crests occurring at the same time) as the
incident photon, the same polarization (direction of oscillation of
the electric field), and is emitted in the same direction as the im-
pinging photon. This situation, depicted in Figure 1.7, is called
stimulated emission. The probability that a photon encountering
a given atom in the appropriate excited state will trigger stimulated emission is the same as that for absorption if the photon were

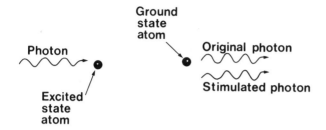

Figure 1.7 Photon incident on excited atom causing stimulated emission.

to encounter this atom in the ground state. The probability of absorption or stimulated emission is much higher than that of spontaneous emission. Since most atoms in a system are normally found in the ground state, absorption is far more likely to occur than stimulated emission. However, if one were to have more atoms in an excited state than in the ground state, i.e. *population inversion,* then an incident photon of appropriate wavelength would trigger stimulated emission, which would result in two photons of equal phase and polarization propagating in the same direction. These two photons could trigger further stimulated emission, which would eventually produce an avalanche of stimulated photons. The resulting light radiation would appear as a well-collimated beam of monochromatic light with a single associated phase and polarization, i.e., we would have *coherent* radiation. In a device which provides such coherent radiation, electrical or optical energy is "pumped" in to achieve population inversion. A photon generated in the active atomic or molecular medium by spontaneous emission and sweeping through the medium is then "amplified" by causing an avalanche of stimulated photons, i.e., we have *l*ight *a*mplification by *s*timulated *e*mission of *r*adiation. Such a device is a *laser.*

1.5 TYPES OF LASERS

Pulsed Lasers

The oldest laser is the ruby laser. In this laser the active medium is a cylindrical rod of ruby, which is an aluminum oxide $(Al_2$

O_3) crystal doped with some 0.05 percent by weight of chromium oxide (Cr_2O_3). Near the ends of the rod, mirrors are placed, one to be totally reflecting, the other to be semitransparent. The medium and these end mirrors form the "laser cavity." The rod is surrounded by a helical flash lamp (pulse duration of a few milliseconds) which optically pumps Cr^{3+} ions to excited states when they absorb blue and green light. The thus excited chromium ions relax to a metastable (long-lived) state. Ions can remain in this metastable state for several milliseconds. If the number of ions in the metastable state is larger than the number of ions remaining in the ground state (population inversion), then a photon, generated by spontaneous emission resulting from relaxation of an ion from the metastable state to the ground state, and travelling through the rod, will trigger stimulated photons. These photons bounce back and forth between the mirrors of the laser cavity to trigger more and more stimulated emission. Since one of the laser cavity mirrors is semitransparent, a laser light pulse exits from the laser cavity. A schematic diagram of a ruby laser is shown in Figure 1.8 and a simplified energy level scheme showing the origin of the red ruby laser light (6943 Å) is presented in Figure 1.9. Since the principle of operation of this laser requires three energy levels (an excited state which is pumped, a metastable state, and the ground state) this laser is called a three-level laser. Other lasers

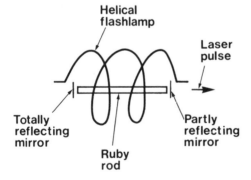

Figure 1.8 Schematic diagram of ruby laser.

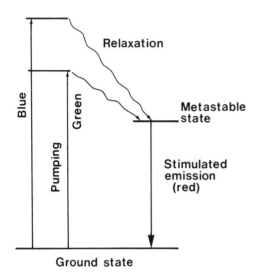

Figure 1.9 Simplified energy level diagram of ruby laser.

operate similarly, but often in a way such that population inversion involves two excited states, rather than one excited state and the ground state. Because four states are now involved in the laser action, such lasers are often called four-level lasers. Figure 1.10 schematically shows the operation of a four-level laser. Pulsed lasers sometimes employ rare-earth ions as the active laser medium. Widely used is neodymium, Nd^{3+}, in glass or yttrium aluminum garnet host materials. The laser wavelength here is 1.06 μm. Instead of solid laser media, gases can be pumped by electric discharge pulses. Excited states are reached by ionization and collisions between molecules or ions. A variety of optically pumped, pulsed lasers utilizing solutions of highly fluorescent dyes as laser media are in existence.

CW Lasers

A number of gas lasers operate in a continuous wave (CW) i.e. not pulsed, fashion. Carbon dioxide lasers, which operate in the infrared (IR), are extremely powerful. Most notable among

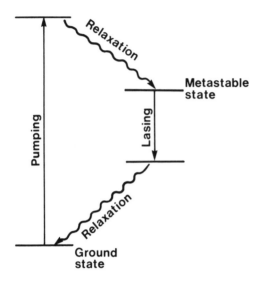

Figure 1.10 Energy level diagram of four-level laser.

CW lasers operating in the visible and ultraviolet (UV) spectral regions are the argon-ion, helium-neon, krypton, and helium-cadmium lasers. For our purposes, the most important of these is the argon-ion laser. In this laser, singly ionized argon is the active laser material. A schematic diagram of an Ar-laser is shown in Figure 1.11. An electric discharge (DC) generates free electrons and argon ions. Since the potential laser levels are high above the ground state, multiple collisions are necessary to raise ions to the required energies. Ar-lasers, therefore, require large pumping currents. Ar-lasers are four-level lasers (see Figure 1.10). With mirrors at both ends of the laser cavity, the Ar-laser will simultaneously "lase" at several lines (colors) in the blue-green spectral region. The main laser lines occur at 5145, 5017, 4965, 4880, 4765, 4727, 4658, and 4579 Å. Single-line operation is achieved by introducing a prism at the cavity end which has the totally reflecting mirror. Light diffraction through the prism bends the various wavelengths to different extent. If light travelling through the prism is

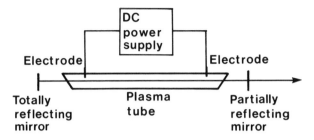

Figure 1.11 Schematic diagram of Ar-laser.

not perpendicularly incident on the totally reflecting end mirror, it will not be reflected back through the laser tube, hence it will not cause stimulated emission, as shown in Figure 1.12. Laser action at the desired line is obtained by suitable rotation of the prism and end mirror. Instead of mirrors designed to reflect blue-green light, namely the normal Ar-laser light, mirrors which transmit visible light but reflect UV light can be substituted, which provides lasing at several lines in the 3340 to 3640 Å region. UV powers of several watts can be achieved with a laser which gives approximately 15 W in normal (blue-green, all lines lasing) operation. In comparison to the powers one is accustomed to deal with in normal lamps, these laser powers do not sound impressive. However, laser beams are well collimated, unlike ordinary light sources which radiate in all directions. Moreover, the laser light consists of a few sharp spectral lines. The laser light one observes is thus extremely bright to the eye. Indeed, precautions are necessary, since the brightness of the laser light can easily blind the observer.

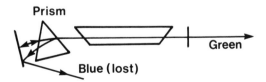

Figure 1.12 Wavelength tuning of Ar-laser.

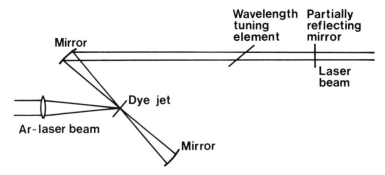

Figure 1.13 Schematic diagram of Ar-laser-pumped dye laser.

It is not advisable to introduce materials such as cloth, wood, paper, etc. into the laser beam, since the beam will often ignite them.

Tunable CW dye lasers are commercially available. These lasers are usually pumped with an Ar-laser or a krypton laser. A schematic diagram of dye laser operation is shown in Figure 1.13. With Ar-laser pumping, lasing throughout the spectral range 530 to 700 nm can be achieved, depending on the choice of utilized dye. For each dye, a continuously tunable range on the order of 50 to 100 nm is obtained. A combination of Ar-laser and dye laser is thus very valuable, making available a wide range of laser wavelengths, from the near-UV through the visible. Figure 1.14 (Plate 1, page 53) shows an Ar-laser/dye laser combination in operation.

REFERENCES

1. B. E. Dalrymple, J. M. Duff, and E. R. Menzel, *J. Forensic Sci. 22,* (1), 106 (1977).

2. J. I. Thornton, *J. Forensic Sci. 23* (3), 536 (1978).

3. E. R. Menzel and J. M. Duff, *J. Forensic Sci. 24* (1), 96 (1979).

4. E. R. Menzel, *J. Forensic Sci. 24* (3), 582 (1979).

5. E. R. Menzel and K. E. Fox, *J. Forensic Sci. 25* (1), 150 (1980).

ADDITIONAL READINGS

Some of the subject matter of Chapter 1 can be found in most first- and second-year college physics or chemistry texts, or in first-year college physical science texts such as

K. Krauskopf and A. Beiser, *Fundamentals of Physical Science*, McGraw-Hill, New York, 1966.

These texts are on an elementary level. On an intermediate level, several texts, such as Eisberg's *Fundamentals of Modern Physics*, Wiley, New York, 1961, can be suggested. Other texts at this level are

G. R. Fowles, *Introduction to Modern Optics*, 2nd ed., Holt, Rinehart and Winston, New York, 1975.

B. A. Lengyel, *Lasers*, 2nd ed., Wiley-Interscience, New York, 1971.

G. Herzberg, *Atomic Spectra and Atomic Structure*, Dover, New York, 1944.

Advanced texts on atomic spectrocscopy are

E. U. Condon and G. H. Shortley, *The Theory of Atomic Spectra*, Cambridge University Press, Cambridge, England, 1964.

M. Tinkham, *Group Theory and Quantum Mechanics*, McGraw-Hill, New York, 1964.

B. G. Wybourne, *Spectroscopic Properties of Rare Earths*, Wiley-Interscience, New York, 1965.

Texts on molecular spectroscopy, though often containing sections which are easily understood, tend to generally be on the advanced level. Such texts are

S. P. McGlynn, T. Azumi, and M. Kinoshita, *Molecular Spectroscopy of the Triplet State*, Prentice-Hall, Englewood Cliffs, 1969.

S. P. McGlynn, L. G. Vanquickenborne, M. Kinoshita, and D. G. Carroll, *Introduction to Applied Quantum Chemistry*, Holt, Rinehart and Winston, New York, 1972.

M. Orchin and H. H. Jaffe, *Symmetry, Orbitals, and Spectra*, Wiley-Interscience, New York, 1971.

F. A. Cotton, *Chemical Applications of Group Theory*, 2nd ed., Wiley-Interscience, New York, 1971.

2 SPECTROSCOPIC INSTRUMENTATION

This chapter describes instrumentation and techniques of optical spectroscopy. Absorption, emission, and emission excitation spectral measurements are emphasized to set the stage for the application of spectroscopic measurements to latent fingerprint detection, treated in Chapter 4. Recall from Chapter 1 that absorption and excitation spectra probe the colors of light which a material can absorb. Within the framework of fingerprint detection by luminescence, absorption or excitation spectra determine the light colors useful for illumination if luminescence is to occur. Comparison of the luminescence spectrum of the material of interest with that of the background, which is usually present, allows selection of filters to maximize contrast in photography of the luminescent latent print.

A variety of commercial instruments specifically designed for absorption, emission, and excitation spectral measurements are on the market. Such instruments (e.g., Cary 17 for absorption, Perkin Elmer MPF-4 for luminescence and excitation) are generally easy to use and are very convenient for routine spectroscopic work. Finances permitting, they are certainly desirable. However, they are not readily adaptable to measurements for which they are not specifically designed and often cannot be used in conjunction with lasers, particularly when fairly large exhibits need to be examined. An exhaustive treatment of spectroscopic instrumentation is clearly beyond the scope of this text. We will therefore concentrate on description of a "modular" spectroscopic system which is such that its individual components can be rearranged and shuffled around to permit several types of spectroscopy. This provides flexibility as well as cost-effectiveness. The latter criterion is important, since spectroscopic equipment is not cheap. While emphasis on a single system may initially seem overly restrictive, the descriptions of the kind of measurement that can be made with it entail the features of a variety of techniques, so that a broad overview of optical spectroscopy emerges.

Sections 2.1 to 2.3 treat the main individual components of the spectroscopic system. The ways in which these components are combined for a variety of measurements are presented in the remainder of the chapter. Section 2.10 list instruments, including manufacturers' names, which the author has found useful.

Spectroscopic instrumentation generally is comprised of three components: a light source which illuminates the sample under study, a monochromator which analyzes the light wavelengths absorbed, emitted, or scattered by the sample, and a light detector with associated electronics for amplification and processing of the detector signal. Experimental arrangements will be shown in block form. Electronic instruments are treated as "black boxes" into which wires (coaxial cables) are plugged to feed in or extract electronic signals. Manipulation of electronic instrumentation is not described in detail, since this is done explicitly in instruction manuals accompanying such equipment. However, the principles of operation of the major constituents of spectroscopic arrangements will be treated.

2.1 MONOCHROMATORS

The central item in a spectroscopic setup is a monochromator. Figure 2.1 is a schematic diagram of the operation of such an instrument. Light is focussed on the instrument's entrance slit and reflected from mirror m_1 such that light rays incident on the diffraction grating are parallel. The grating separates colors, as shown in Figure 2.2. Light reflected from the grating is focussed by mirror m_2 (Figure 2.1) on the exit slit. The color of light reaching the exit slit is determined by the angle of rotation, a, of the grating. Wavelength scanning is achieved by an electrically or man-

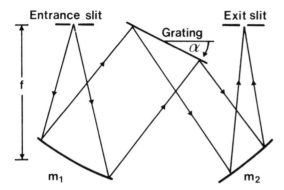

Figure 2.1   Schematic diagram of monochromator (Czerny-Turner type). f = focal length; α = angle of rotation of the diffraction grating; m_1, m_2 = mirrors.

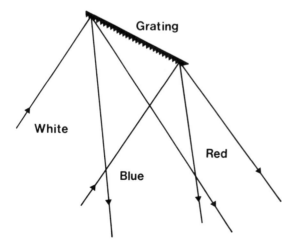

Figure 2.2   Wavelength separation by a diffraction grating.

ually driven precision screw which is connected to the grating. The wavelength resolution of the instrument depends on several factors: the focal length, f (see Figure 2.1), the number of lines per inch cut into the grating, and the widths of the entrance and exit slits. The larger the focal length and the number of lines per inch of the grating and the narrower the slits, the better the resolution. For our purposes, resolution is not a critical requirement, since even small monochromators (f = ¼ m) have resolutions of the order of 1 Å, higher than usually necessary in molecular spectroscopy of the kind applicable to fingerprint luminescence. In fact, shorter focal length leads to greater instrument light throughput hence greater sensitivity. However, short focal length monochromators tend to suffer from considerable stray light throughput by scattering from the walls of the instrument, which can be a problem. Often the use of filters at the monochromator entrance slit can remedy this situation.

2.2 PHOTOMULTIPLIER TUBES

The light reaching the monochromator exit slit is detected by use of a photomultiplier tube. This is a vacuum tube equipped with a transparent window, a photosensitive material onto which the light falls, and a series of dynodes. A schematic diagram is shown in Figure 2.3. Photons incident on the photocathode (the photosensitive material) cause it to eject electrons. These electrons are

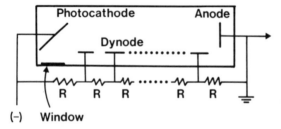

Figure 2.3 Schematic diagram of photomultiplier tube. R = resistor. A high (−) voltage is applied between the photocathode and the anode.

electrostatically attracted to the first dynode. This is accomplished by connecting the resistor chain of Figure 2.3 to a DC power supply, thus applying a voltage between the photocathode and dynodes. Voltages are typically 900 to 1500, depending on the type of photomultiplier used. The high-voltage DC power supply is not explicitly shown in the block diagrams of the chapter. An electron striking the first dynode causes it to eject several secondary electrons (typically three). These electrons are electrostatically attracted to the second dynode, where, again, each incident electron creates some three secondary electrons. Photomultipliers usually have 9 to 13 dynodes, so that a primary electron ejected from the photocathode results in an avalanche of some 10^4 to 10^6 electrons collected at the anode. The output from the photomultiplier (anode) is fed to a suitable electronic black box which further amplifies and processes the signal. The spectral sensitivity of photomultipliers depends on the material of the photocathode. Sensitivities range from the UV to the near IR (about 1.1 μm). Figure 2.4 shows the spectral sensitivities of photomultipliers of S-1 and and S-20 response as examples. The designations S-1 and S-20 are manufacturers' codes denoting the photosensitive material of the photocathode. The percent quantum efficiency in Figure 2.4 means the percent probability that one photon striking the photocathode will cause it to eject one electron. The overall sensitivity of a photomultiplier depends not only on the photocathode material, but also on the number of stages (dynodes). Photomultipliers can have substantial dark current, i.e., signal output in absence of incident light. This dark current arises from electrons ejected thermally from the photocathode or dynodes. For nine-stage photomultipliers of S-20 or similar response, this dark current is, for most purposes, not important. However, in 11-stage tubes of S-1 response this dark current can be quite large, so much so that it can interfere with detection of weak fluorescence, for instance. The dark current can be reduced dramatically (by some four orders of magnitude) when the photomultiplier is cooled. Commercial cooling housings which use a dry ice/methanol coolant mixture achieving –70°C do this, and are almost mandatory with S-1 photomultiplier tubes.

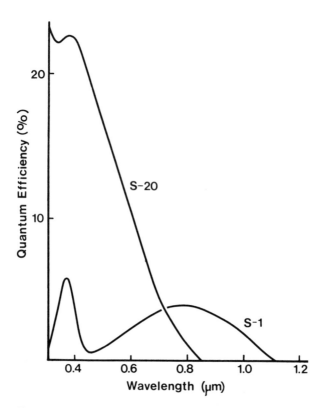

Figure 2.4 Quantum efficiencies of S-1 and S-20 photocathodes.

2.3 DETECTION ELECTRONICS

While there are a variety of instruments, such as lock-in amplifiers, boxcar averagers, waveform eductors, picoammeters, etc., suitable for signal processing, we will concentrate on the description of photon counting instrumentation in keeping with the intent of description of a flexible modular system. Photon counting instrumentation is sensitive, cost-effective, and can perform a large variety of functions. The current output from the photomultiplier is schematically shown in Figure 2.5. The noise (dark current) has a number of superimposed spikes, each of which corresponds to a photon which has struck the photcathode to eject a photoelectron. This output is fed to an amplifier/discriminator. This black box amplifies the signal and rejects the noise (spikes smaller than a predetermined height). It further shapes the spikes so that all output spikes have equal height and width, as shown in Figure 2.5. The output from the amplifier/discriminator is fed to a photon counter/processor where the spikes are counted up for an adjustable period of time and the number of counts is displayed digitally. Since thermal emission of electrons from the photocathode cannot be discriminated against by the amplifier/discriminator (whereas electrons thermally ejected from a dynode, particularly one toward the end of the chain, are rejected), the photon counter/processor may display a non-zero count in absence of light, which can be significant. Photon counters, therefore, have a background subtraction feature, whereby the background is automatically subtracted from the number of counts obtained in presence of light. Some photon counters can be gated. This is a valuable feature. In this mode of operation, a periodic voltage pulse triggers the instrument. A gate then counts for an adjustable period of time. The onset of the gate can be delayed in time with respect to the triggering pulse as well. The gated mode of operation is shown in Figure 2.6. It is sometimes convenient to illuminate a sample not continuously, but in a chopped fashion, wherein the light source is periodically interrupted by a light chopper if one does not use a pulsed light source. A light chopper may be as simple as a motor-driven wheel with slots in it (mechanical light chopper). Commer-

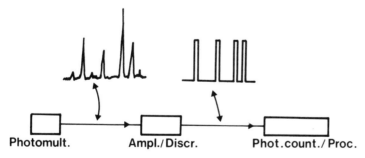

Figure 2.5 Principle of photon counting system (see text).

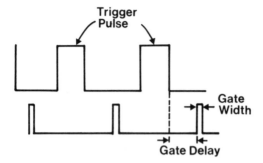

Figure 2.6 Gated photon counting. A triggering pulse activates the photon counter/processor which accumulates signals starting at a specific (adjustable) time, the gate delay, after the triggering pulse and for a specific (adjustable) interval, the gate width.

cial mechanical light choppers are equipped with circuitry which provides a periodic electronic signal suitable to trigger a photon counter. Electrooptic modulators do the same job, but they are expensive and have functions which far exceed the needs of the present text. While the gated photon counting feature is perhaps not essential, it is nonetheless convenient, as it allows a variety of measurements, such as luminescence lifetime work in the range of micro- or milliseconds (or longer), to be made. A further, very desirable, feature of some photon counting systems is the two-channel feature which allows the signals from two separate amplifier/discriminators to be fed to a single processor. Counts are accumulated in two separate channels. The difference or sum of these counts, or the counts from each channel, can then be displayed digitally. An application of the two-channel feature will be given in Chapter 4. The photon counter/processor should have an analog output so that counts can be monitored by charter recorder. If one operates in a gated mode, an oscilloscope should be on hand to allow monitoring of the triggering signal as well as gate positions and widths. The oscilloscope should have a dual-trace plug-in to allow simultaneous monitoring of two separate inputs to the oscilloscope, but otherwise need not have any sophisticated features.

2.4 ABSORPTION SPECTRA

Figure 2.7 is a schematic diagram of an absorption spectrophotometer. The output from a broad-band light source, such as a tung-
sten lamp, illuminates a monochromator. The output from the
monochromator is split into two beams. One of these illuminates
a sample under study; the other illuminates a reference sample. In
solution spectroscopy, for instance, the sample under study is a
solute contained in a solvent and the reference is solvent only. The
amount of light transmitted through the two sample cells falls onto two photomultipliers. The detection electronics, such as a two-
channel photon counter, subtracts the signal of the sample photomultiplier from that of the reference photomultiplier. The difference signal is then amplified and monitored by a chart recorder.
The absorption spectrum of the sample is obtained by wavelength
scanning of the monochromator. Sometimes a light chopper with a
reflecting slotted wheel is incorporated into the system instead of
the beam splitter, so that light falls alternatively on the sample and
reference cells. Now only a single photomultiplier is needed, as
shown in Figure 2.8. The detection electronics is now phase and
frequency selective.

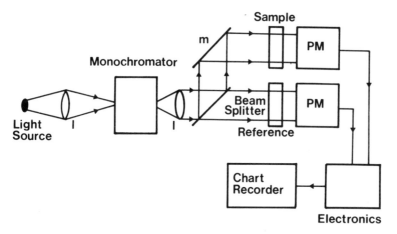

Figure 2.7   Schematic diagram of absorption spectrophotometer. m = mirror,
l = lens; PM = photomultiplier tube.

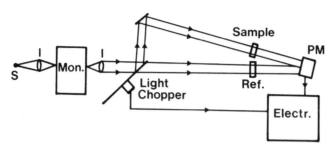

Figure 2.8 Schematic diagram of an absorption spectrophotometer using AC detection electronics. Mon. = monochromator; S = light source.

A simpler experimental arrangement, which is part of the modular system to be emphasized, is shown in Figure 2.9. Here, the broad-band light source illuminates the sample under scrutiny. Light transmitted by the sample falls onto a monochromator which is wavelength scanned. Once the absorption spectrum is obtained, a reference cell is substituted, if necessary, and a spectrum is again obtained. The subtraction of one spectrum from the other is done manually to yield the corrected absorption spectrum. While the manual subraction and the need to run two separate spectra make this procedure a bit more laborious than the earlier configurations, the system is a simple one, which can be converted readily to luminescence and excitation spectral measurements. In

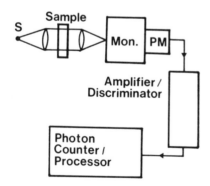

Figure 2.9 Absorption spectrophotometer using photon counting.

solution-absorption spectroscopy, the solvent is often transparent in the wavelength region of interest, and a reference spectrum is then not required.

2.5 WAVELENGTH RESPONSE CALIBRATION

Because the efficiency with which diffraction gratings reflect light is wavelength dependent and because the sensitivity of photomultiplier photcathodes is wavelength dependent, the spectral response of the system (Figure 2.9) is not uniform. Diffraction gratings can have various "blaze" wavelengths. The blaze gives the wavelength at which the grating performs best. If the grating blaze and the photomultiplier response are suitably selected, then the nonuniformity in spectral response can be minimized. Nonuniformity in spectral response versus wavelength of light is, in any event, not always critical. However, when uniformity is required to obtain true, rather than distorted, spectra, then the instrumentation must be calibrated (this is not just a feature of the present configuration; commercial instruments come factory calibrated). The determination of the wavelength response of the system can be made in several ways. If two monochromators and a wavelength-calibrated light power meter (such meters are commercially available) are on hand, then the system calibration is fairly straightforward. The need of a second monochromator for calibration purposes does not imply additional expense or complication, because the system which will be suggested in Chapter 4 as a highly useful one for forensic analysis contains two monochromators. The broadband light source (lamp) illuminates one of the monochromators. The output of the monochromator is measured as a function of wavelength with the power meter. This light output, now of known intensity as a function of wavelength, serves as input to the second monochromator, which has a photomultiplier at the exit slit. The wavelength of the second monochromator is then scanned, with the light output from the first monochromator tracking wavelength. The thus obtained spectrum, recorded by detection electronics associated with the second monochromator, is then compared with the known light input into the second monochromator. This comparison yields the spectral correction factors necessary to calibrate the monochromator/photomultiplier combination. As we shall see, the most useful spectroscopy for forensic purposes will be luminescence spectroscopy using an Ar-laser for sample illumination. It can also be anticipated that a CW dye laser, pumped by the CW Ar-laser, will eventually be useful. If one has an Ar/dye laser system on hand, then wavelength calibration can be made over the range 450 to 700 nm using laser wavelengths. In this case a second monochromator is not required. Laser wavelengths (attenuated by neutral density filters to avoid damaging the photomultiplier), whose power is measured (one usually acquires a power meter when purchasing a CW laser), serve as input to the monochromator/photomultiplier combination to be calibrated. If, for instance, blue light and red light of equal intensity fall onto the monochromator entrance slit, but the system is twice as sensitive in the blue as in the red, then the chart recorder will record twice the intensity in the blue as in the red. The relative spectral correction (absolute calibration is not necessary) is made simply by multiplying "red signals" by a factor two with respect to blue signals to obtain a true spectrum.

For measurement of excitation spectra, it is necessary to know the wavelength response of a monochromator alone (no photomultiplier at the monochromator exit slit) in order to obtain undistorted spectra. The output from one monochromator (or, again, laser lines may be used) is measured with a power meter, and then serves as input to the second monochromator, which is to be calibrated. The output from the exit slit of this second monochromator is then measured with the power meter and this provides the wavelength response.

2.6 EMISSION SPECTRA

Having absorbed light at a given wavelength, a number of materials re-emit light at a longer wavelength (fluorescence or phosphorescence), as described in Chapter 1. Figure 2.10 shows an experimental arrangement to measure luminescence spectra. Light, here

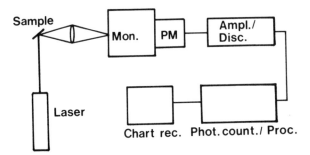

Figure 2.10 Experimental arrangement for luminescence spectroscopy using sample excitation by laser.

from a suitable laser such as an Ar-laser, is incident on the sample under scrutiny. Luminescence from the sample, which is radiated in all directions, is focussed with a lens on the entrance slit of a monochromator. Some of the laser light is scattered and can also reach the monochromator entrance slit. This light is not wanted but, because the monochromator is set at wavelengths longer than the laser wavelength in order to detect luminescence, the laser light reaching the monochromator is rejected by it. If the mono-chromator has short focal length, such that significant stray light through-scattering can occur, a long-wavelength-pass filter is plac-ed near the monochromator entrance slit to block the laser light. It should be noted that, except for the fact that the laser illumi-nates the sample at right angle with respect to the optical path to the monochromator (the angle here is not important), there is essentially no difference between the luminescence setup and the absorption arrangement of Figure 2.9. Thus, one spectroscopic system serves two purposes. The luminescence from the sample is passed by the monochromator which can be wavelength scanned to yield the luminescence spectrum.

It is frequently found that a sample not only shows lumines-cence from the material of interest, but background luminescence as well. It may be necessary, or at least desirable, to subtract out this background. The procedure to do so will be given in example form. Suppose we have a latent fingerprint which luminesces on a surface which also luminesces. First a region of the surface containing no fingerprint is illuminated and the background emission spectrum is measured. Then the fingerprint region is illuminated and the spectrum, which now is a superposition of fingerprint and background emission, is obtained. In the region containing the fingerprint, some of the illuminating light which would otherwise reach the substrate to cause background emission may be absorbed and/or scattered by the fingerprint residue. Also, substrate emission may be lost by scattering from the fingerprint residue. The background luminescence content in the spectrum taken from the fingerprint region may therefore be significantly lower than the spectral intensity obtained from the bare surface region. The bare substrate spectrum is therefore corrected as shown in Figure 2.11 and subtracted from the print + background spectrum. In the example of this figure, the latent print luminescence is shown as occurring between 520 and 620 nm. The background emission from a region not containing a latent print is compared with the print + background spectrum at a wavelength at which there is no fingerprint luminescence. For instance, 650 nm is such a wavelength in Figure 2.11. Generally, a wavelength longer than the fingerprint luminescence wavelengths is chosen. Choice of this wavelength can usually be made by inspection of the background and print + background spectra. In Figure 2.11, the background spectrum at 650 nm is some 25 percent higher than the print + background

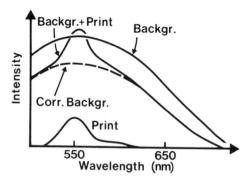

Figure 2.11 Schematic representation of procedure for background subtraction in emission spectra (see text).

spectrum. The background spectral intensity is therefore corrected
by a multiplicative factor of 0.75 at all wavelengths, which yields
the spectrum shown as the dashed line in Figure 2.11. This spectrum is subtracted from the print + background spectrum to yield
the print spectrum. Rather than using algebraic correction and
manual plotting, the background correction can be done experimentally. The background and print + background spectra are first
taken. A suitable wavelength of no fingerprint luminescence is
next chosen. The monochromator is then set at this wavelength
and the chart recorder pen is set at the appropriate point on the
already recorded print + background spectrum. A bare portion of
the substrate is next illuminated and the background intensity is
adjusted (by changing monochromator slit widths, reducing illumination intensity, changing photon counter accumulation time,
etc.) such that the recorder pen position is superposed on the print
+ background trace. The background spectrum is then run with
wavelength scan and chart paper directions reversed. In terms of
Figure 2.11, the chart recorder traces would be the background +
print and the corrected background lines. A more elegant experimental approach to background subtraction for purposes of latent
fingerprint detection will be described in Chapter 4. If necessary,
instrumental wavelength response calibration is made in the manner
of the previous section.

It is sometimes of interest to determine not only the color of
luminescence (i.e., the luminescence spectrum) of a given sample,
but also the strength of this luminescence. The luminescence
strength, termed luminescence quantum efficiency (or quantum
yield), gives a measure of the probability that one directly absorbed photon will create a luminescent photon. Quantum yields are
therefore at most equal to one, a case almost never found. Laser
dyes have high fluorescence yields (0.3 to 0.9), whereas amor-
phous solids at room temperature usually fluoresce only weakly
(yields 10^{-7} to 10^{-3}), if at all. The optical density of the sample
(see Chapter 1) gives a measure of the number of incident photons
absorbed. The luminescence intensity gives a measure of the number of photons emitted. Once these two quantities have been measured, comparison with a standard of known luminescence yield

can provide the luminescence efficiency of the sample under study. A number of alternative methods for quantum yield determination exist, some of which do not require a standard. However, precise quantum yield determinations are generally quite difficult, and will not be treated in this text. The interested reader is referred to an excellent review by Demas and Crosby [1]. For purposes of laser detection of latent prints, quantum yield determinations are usually not necessary, except, perhaps, in research work. If, for instance, one wishes to compare the sensitivities of two compounds with which exhibits are treated (see Chapter 3), it generally suffices to compare their luminescence intensities at the wavelengths of maximum luminescence (with compounds deposited on exhibits to comparable optical density) to arrive at an estimate of their relative merits.

2.7 EXCITATION SPECTRA

In some instances it may be convenient to determine the excitation (illumination) wavelength which optimizes the emission of interest. Alternately, if luminescence is observed from a sample, it may be of interest to obtain the absorption spectrum of the compound which shows the observed luminescence. Excitation spectra provide this information, particularly in the case of samples containing several species, only one of which luminesces. Figure 2.12 shows the appropriate experimental arrangement. Monochromator 1 provides the excitation to the sample and is wavelength scanned. Monochromator 2 is set at the luminescence wavelength and is not wavelength scanned. The luminescence intensity is thus monitored as a function of sample excitation wavelength. The obtained intensity spectrum corresponds to the absorption spectrum of the luminescent species because fluorescence occurs only from the lowest level of the first excited singlet state as a result of rapid radiationless relaxation among excited states (see Chapter 1). Consequently, the luminescence intensity (fluorescence or phosphorescence) is proportional to the amount of light absorption into any excited singlet state. In a sense, absorption and excitation spectra provide the same information. However, absorption spectra cannot be

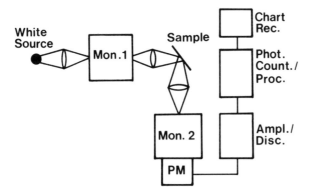

Figure 2.12   Experimental arrangement for excitation spectroscopy.

measured on opaque samples, but excitation spectra often can. In systems containing several species, absorption spectra are only of limited use. The experimental arrangement shown in Figure 2.12 is identical to those of Figures. 2.9 and 2.10 with respect to the monochromator/photomultiplier and detection electronics part. Thus, in essence, the system performs three tasks.

2.8 RAMAN SPECTRA

While Raman spectroscopy will only be mentioned briefly in the remainder of this text, the experimental technique is so similar to fluorescence techniques that a short description is provided with potential future utility in forensic work in mind.

   Just as a billiard ball bounces off a cushion with loss of energy, a photo incident on a molecule can be scattered by it with loss of energy. The scattered photon thus changes color (to longer wavelength). The energy lost by the photon is taken up by the molecule through change in vibrational state. Raman spectra, in a manner analogous to infrared spectra, thus probe the vibrational structure of molecules. Raman spectra do not, however, duplicate IR spectral information, since vibrations accessible to Raman spectroscopy are usually not accessible to IR spectroscopy, and conversely. Raman spectroscopy is singularly well suited to the use of lasers (because of their high power and monochromaticity), a second reason for the present description. Experimentally, Raman spectra are obtained much like luminescence spectra. The experimental arrangement is none other than that shown in Figure 2.10. The exciting laser wavelength, however, is chosen not to fall into a region in which the molecule of interest absorbs. This choice is made because Raman spectra are weak and can be obscured by fluorescence. Since stray light scattering through the monochromator can be very detrimental, a filter blocking laser light is almost mandatory at the monochromator entrance slit. If one has several laser lines on hand, as in a CW Ar-laser, a Raman spectrum can be quite easily distinguished from a fluorescence spectrum. Raman lines occur at fixed energies lower than the laser photon energy. Thus, if the laser wavelength is changed, the Raman spectrum will also shift in wavelength. Fluorescence, on the other hand, does not shift in wavelength (but may, of course, change in intensity).

2.9 LUMINESCENCE LIFETIME MEASUREMENT

From a practical standpoint, the distinction between fluorescence and phosphorescence is made on the basis of lifetime, as mentioned in Chapter 1. We will encounter both fluorescence and phosphorescence detection of latent fingerprints in Chapter 3. The determination of the nature of the observed luminescence with respect to its lifetime is, therefore, briefly considered. We restrict our attention primarily to methods which utilize instrumentation similar to that already described. As mentioned earlier, luminescence lifetimes in the millisecond and microsecond time domain can be measured with a photon counting system operating in the gated mode.

Let us first consider the meaning of a luminescence lifetime. If a light source illuminating a luminescent sample is suddenly shut off, the luminescence intensity will decrease as a function of time after illumination cessation in accord with the relation

$$\frac{I}{I_0} = e^{-t/\tau}$$

$$(2.1)$$

where the number e has a value of about 2.7. I_0 is the luminescence intensity just prior to illumination cutoff and I is the luminescence intensity at time t after cutoff. The quantity τ is the luminescence lifetime. The luminescence intensity decrease is exponential in time. When the intensity has decreased to about 0.37 of the initial intensity, then one lifetime has elapsed. We note that the definition given here is different from that of a half-life, which corresponds to the time lapse necessary for reduction to 0.5 of the intial intensity.

Figure 2.13 shows an experimental arrangement for lifetime measurements using gated photon counting. A mechanical light chopper provides the light cutoff and a signal to trigger the photon counter which operates in the gated mode. The photon counter gate is positioned as shown in Figure 2.6 at a time t after light cutoff. Once the luminescence intensity is measured, the gate is moved to a new time after cutoff and intensity is measured again. This procedure is repeated as often as needed. Intensity versus delay time is then plotted and compared to relation 2.1 to yield τ.

An instrument which performs in a manner similar to the gated photon counter is a boxcar averager. This instrument replaces the amplifier/discriminator and photon counter/processor in Figure 2.13 when used for lifetime work. This instrument, too, is triggered by the light chopper in Figure 2.13 (or an electrooptic modulator for measurement of short lifetimes which are beyond

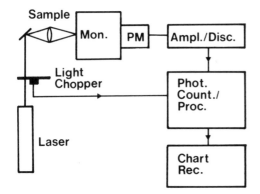

Figure 2.13 Emission spectroscopic arrangement with gated photon counting (see text).

the capabilities of mechanical choppers). It also has a gate which measures intensity at a time after light cutoff. The gate, however, can automatically scan times after cutoff, rather than needing to be changed manually. Rather than photon counting principles, the boxcar averager uses frequency-selective signal averaging, wherein signals are averaged over many cycles, instead of being counted up over a fixed, large number of cycles. Boxcar averagers are not nearly as intensity sensitive as photon counters, but they permit measurement of shorter lifetimes, sometimes as short as 10^{-9} sec. When dealing with fluorescence lifetimes in solids, one finds that they often fall into a time domain even shorter than nanoseconds (10^{-9} sec), namely the picosecond (10^{-12} sec) domain. A number of techniques to perform lifetime measurements on this time scale have been developed [2-4]. These techniques are very complex, however, and, in most instances, utilize instrumentation quite different from that emphasized in this text. In view of the recognition that forensic laboratories generally will not wish to expend large sums to acquire highly sophisticated instrumentation of the type necessary for picosecond spectroscopy, we forego description of picosecond spectroscopic techniques in this text. The reader is, however, encouraged to peruse the picosecond spectroscopy literature, since picosecond techniques have already found exciting applications in many fields, and may possibly be of eventual potential in forensic analysis also.

2.10 INSTRUMENTS

This section presents a list of equipment the author has found useful in spectroscopic work addressing latent fingerprint detection and study of fingerprint luminescence features. Names of manufacturers are given. The equipment list is restricted to instrumentation in line with the emphasis of this chapter. For a more complete listing of instrumentation and suppliers, the annual buyers' guide of *Laser Focus*, a magazine dedicated to the laser field, and the annual vendor selection issue of *EOSD*, a magazine of electrooptic and laser technology, are good references.

Item	Manufacturer
CW Ar-laser with about 15 W power (all lines visible), equipped with UV option, CW dye laser	Spectra-Physics; Coherent Radiation
Two 1/4 or 0.3 in monochromators	Spex; Jarrell Ash; McPherson
Two photomultipliers of equal spectral response; for most purposes, 9-stage S-20 tubes should be adequate	RCA; EMI; Amperex
DC high-voltage power supply delivering 0 to 2000 V	Fluke; Gencom
Photon counting system, two channel, gated, with two amplifier/discriminators	Princeton Applied Research (Model 1112 processor and Model 1120 discriminators)
Chart recorder	Brinkmann; Hewlett Packard
Light power meter	Karl Lambrecht
Oscilloscope	Tektronix; Hewlett Packard
Mechanical light chopper	Princeton Applied Research
Lenses, mirrors, neutral density filters, beam splitters, lens holders, optical benches, etc.	Oriel; Ealing; Cenco
Long-wavelength-pass filters with cutoff wavelengths ranging from 400 to 700 nm	Corning; Oriel; Ealing
Band-pass interference filters with band width 50 to 100 nm; with band width 10 to 20 nm	Ditric Optics; Oriel
Laser safety goggles	Fisher

The rationale for listing several types of filter will become clear in Chapters 3 and 4, where their essential role in latent fingerprint detection by laser is treated.

REFERENCES

1. J. N. Demas and G. A. Crosby, *J. Phys. Chem. 78*(8), 991 (1971).
2. R. R. Alfano and S. L. Shapiro, *Physics Today* (July, 1975).
3. C. V. Shank and E. P. Ippen, *Laser Focus* (July, 1977).
4. E. R. Menzel and Z. D. Popovic, *Rev. Sci. Instr. 49*, 39 (1978).

ADDITIONAL READINGS

The following book contains spectroscopic data on a variety of compounds, including biologically important species which may be relevant to forensic analysis. Spectroscopic instrumentation and techniques are described as well.

G. G. Guilbault, *Practical Fluorescence, Theory, Methods, and Techniques,* Marcel Dekker, New York, 1973.

This text presents spectroscopic theory as well as techniques.

A. A. Lamola (Ed.), *Creation and Detection of the Excited State,* Vol. 1, Pt A, Marcel Dekker, New York, 1971.

This book presents absorption and emission spectra of a large number of organic molecules. Molar extinction coefficients and lifetimes of fluorescence and phosphorescence are also given for a number of compounds.

I. B. Berlman, *Handbook of Fluorescence Spectra of Aromatic Molecules,* 2nd ed., Academic, New York, 1971.

3 LASER DETECTION OF LATENT FINGERPRINTS

In this chapter, the detection of latent fingerprints by laser is treated. Section 3.1 reviews some conventional fingerprint development methods as well as two recently devised techniques which are still in the research stage. In Section 3.2, laser detection through inherent fingerprint luminescence is described. Section 3.3 deals with laser detection in conjunction with a number of treatments of latent prints with fluorescent materials, while Section 3.4 describes the use of several chemicals which react with fingerprint residue to form luminescent reaction products amenable to laser detection. Latent print treatment with phosphorescers is described in Section 3.5. Finally, examples of application to criminal cases are given in Section 3.6.

Examination of the skin of fingers under microscope reveals a series of pores spaced along the friction ridges which cover the skin. A portion of the fluid excreted through these pores is left on surfaces when they are touched, duplicating the friction ridge pattern thus forming a latent print. A typical latent fingerprint deposit weighs perhaps a tenth of a milligram. Some 99 percent of this deposit is water, which soon evaporates to leave a residue comprised of roughly equal amounts of inorganic material, such as salt, and organic components, including amino acids, lipids (i. e., oils, fats, waxes), and vitamins [1]. Since the amount of material in the latent print residue is extremely small, latent prints are usually not observable by simple inspection of touched surfaces. Latent print examiners, therefore, use a variety of materials for treatment of latent prints to render them visible.

3.1 SOME CONVENTIONAL METHODS

The conventional methods for detection of latent prints exploit either a physical or a chemical property of the fingerprint residue. Dusting, perhaps still the most frequently used method for latent print development, relies on the adherence of fine powder to the residue comprising the latent print. Dusting powders are often composed of two main ingredients. One of these is a material whose function it is to provide good adhesion, such as a resinous polymer. This material is used as a binder into which a colorant is incorporated. The colorant provides contrast between the dusted print and the substrate, necessary for inspection and subsequent photography or lifting. Iodine fuming exploits a similar property, namely the absorption of iodine by lipids of the fingerprint deposit, so that a visualizable brown stain is obtained. The silver-nitrate treatment relies on a chemical property of the fingerprint residue, namely the reaction of silver ion with chloride ion of the salt in the fingerprint deposit. Exposure to light following the reaction, much like silver-halide photography, leads to reduction of silver-ion to silver, whereby a dark "stain" is formed. In the early 1950s, the ninhydrin method was discovered. It relies on the reaction of ninhydrin with amino acids of the fingerprint deposit, which leads to the formation of a purple reaction product. The above methods, as well as others, such as the iodine-silver plate method, provide the forensic analyst with a considerable range of fingerprint detection avenues. Nonetheless, there are frequent cases in which latent prints are difficult to detect by these methods. Dusting, for instance, is only effective for a limited time after fingerprint deposition. Once the deposit dries out, it no longer effectively holds dusting powders. Chemical methods, which generally yield dark stains, are ineffective on many dark surfaces because of lack of contrast. The silver-nitrate treatment suffers if the latent print has been exposed to moisture because of bleeding or removal of salt from the deposit. It would be valuable to have procedures which can be effective in such situations, and vigorous research of new techniques is therefore still ongoing.

Several exciting new methods have been developed in recent

years. Particularly noteworthy among these are metal evaporation in vacuo and autoradiography. The former technique relies on the inhibiting effect of fingerprint residue on the ability of metals to adhere to surfaces. The surface suspected to contain a latent print is inserted into a vacuum chamber, and thin films of metal are evaporated onto the surface. The fingerprint ridge detail is revealed by absence of metal deposition at the latent print ridge site. A combination of gold and cadmium appears to be the best combination of metals found to date. The method has been reviewed by Thomas [2]. In autoradiography, the latent print is labeled with a radioisotope. The latent print, which thus becomes radioactive, is brought in contact with a photographic emulsion, usually x-ray film. The labeling of latent prints can be accomplished by reagents containing radioisotopes, such as formaldehyde or stearic acid (containing ^{14}C), sulfur dioxide or thiourea (containing ^{35}S), silver nitrate (containing ^{110}Ag), or radioactive halogens (^{131}I or ^{82}Br). Alternatively, in situ labeling through neutron activation, producing radioactive ^{24}Na or ^{38}Cl, has been explored. The autoradiography method has been reviewed by Knowles [3]. These methods are sophisticated. Perhaps for that reason they are not routine forensic methods at this time. Also, they have several limitations, some coinciding with those of conventional procedures.

A reasonably straightforward method, applicable to most situations, would therefore be desirable. More than that, it would be useful to have a detection method which does not necessitate treatment with chemicals, including dusting, which might interfere with other exhibit examination procedures for, as an example, blood analysis.

3.2 LASER DETECTION OF LATENT PRINTS— INHERENT FINGERPRINT LUMINESCENCE

One might consider the optical properties of fingerprint residue itself as possible avenues for fingerprint detection. While fingerprint residue contains a variety of compounds which absorb in the visible and near-UV spectral regions, light absorption, or for that

matter, scattering, is generally not effective because of the very
minute quantities of material in the fingerprint deposit. Fingerprint residue contains known luminescers, such as riboflavin and
pyridoxin [4]. Even though spectroscopic techniques can detect
trace amounts of luminescent material with great sensitivity, it has
been found that illumination of latent fingerprints with ordinary
light sources, followed by visual inspection to detect fingerprint
luminescence, is only rarely effective [5]. This lack of success has
two reasons. Ordinary light sources radiate not only at wavelengths at which a luminescer absorbs, hence at wavelengths which
can cause luminescence, but at a broad range of nonabsorbed
wavelengths as well. Only a small portion of the lamp spectrum is
therefore useful, and most of the lamp power is wasted. A portion
of the lamp spectrum also falls in the wavelength range in which
the material luminesces. This portion of the lamp spectrum has to
be *scrupulously* filtered out prior to light incidence on the exhibit.
Otherwise the incident light intensity completely overwhelms the
possible fingerprint luminescence which can be expected to be
quite weak. This filtering presents formidable problems. Not even
relatively small amounts of stray light at the luminescence wavelength can be tolerated because the amount of luminescer in the
fingerprint deposit is only on the order of 10^{-9} grams. Unfortunately, during the course of the required careful filtering, the useful
portion of the lamp spectrum is also drastically reduced, so that
the useful lamp light is actually quite feeble. A lot of lamp light
is wasted, in any event, because lamps radiate in all directions and
only a portion of this light can be collected by mirrors or lenses in
order to illuminate the exhibit under scrutiny.

    One is thus drawn to consider laser for possible exhibit illum-
ination. Lasers which radiate at a wavelength at which a luminescer absorbs are uniquely well suited for luminescence detection
because none of the above-cited problems of ordinary lamps per-
tain. Laser light is monochromatic (or comprised of a small group
of well-defined wavelengths), so that light filtering prior to incidence on the exhibit is not required. The laser beam is well collimated so that no light is wasted. It should be noted that a given
laser may only radiate a few watts, which does not seem power-

ful. However, because of the collimated and monochromatic nature of the laser beam, this amount of light is literally blindingly
powerful. Indeed, it can actually burn exhibits if the beam is not
expanded to illuminate a much larger area than a directly incident
beam.

It turns out, as shown by spectra in Chapter 4, that CW Ar-
lasers are very well suited for the detection of latent fingerprints.
Pulsed lasers which might be suitable wavelength-wise tend to have
low pulse repetition rates, which makes exhibit inspection quite
cumbersome. Even though the power of each laser pulse can be
large, pulse durations are usually short (about 10^{-8} sec for
nitrogen and neodymium: glass lasers, for instance) so that the
laser power over a time span of seconds or minutes is generally
low in comparison with Ar-lasers. Time spans of seconds and minutes are of interest in photography of latent print luminescence
under laser light, as will be described shortly. Blue-green laser
wavelengths turn out to be very suitable for fingerprint detection.
Near-UV laser wavelengths have valuable application as well. These
wavelengths are available with Ar-lasers. Indeed, they are precisely
the Ar-laser wavelengths. The power of Ar-lasers is considerably
higher than that of krypton or helium-cadmium lasers which have
laser lines of suitable wavelengths. Should forensic applications
eventually require wavelengths longer than those of an Ar-laser,
one can always couple the Ar-laser to a CW dye laser to achieve
such wavelengths, with rather high laser power obtainable.

The laser procedure for detection of latent fingerprints
through inherent fingerprint luminescence takes on the following
simple form [5]. The exhibit under scrutiny is illuminated with
the blue-green light beam from a CW Ar-laser. The laser beam is
expanded with a lens (convergent or divergent) to illuminate an
area large enought to cover a whole fingerprint, or larger. Fiber
optics can be used as well, and provide ease of laser beam manipulation. In addition to allowing inspection of a complete finger-
pring, the beam expansion protects the exhibit from overheating,
which can lead to burning. The Ar-laser usually operates at blue-
green wavelengths (all lines lasing). Sometimes better contrast is
obtained for latent prints on substrates such as certain types of

paper if the Ar-laser operates at 5145 Å rather than at all lines. The *very bright* laser light scattered or reflected from the illuminated exhibit can cause severe eye damage. Laser safety filters, such as Fisher 11-409-50A, which absorb the laser wavelengths and transmit at wavelengths longer than roughly 540 nm—thus transmitting at least part of the fingerprint luminescence, which is greenish yellow in color, sometimes orange, namely in old prints—are therefore used to protect the observer's eyes during exhibit inspection. The necessity of safety goggles cannot be overstressed. Eye damage can result from even relatively low laser intensities (see Chapter 5). Once a fingerprint is located, it is photographed. Inspection and photography are, of course, carried out in a darkened room. The photography utilizes the same safety filter so that only luminescence reaches the film. The procedure is schematically shown in Figure 3.1. The film used for photography should be reasonably fast, since the luminescence from latent prints is usually quite weak. Photographic exposures of seconds and sometimes minutes are generally needed. Indeed, careful inspection of exhibits is usually necessary in order not to overlook the often only very faintly luminescent ridge detail. The film should have fairly high contrast, since background luminescence from the substrate is almost always present. Indeed, in addition to the above safety filter, narrow band-pass filters are sometimes needed to enhance contrast sufficiently to allow photography. Filters will be

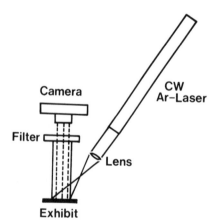

Figure 3.1 Apparatus for detection of latent prints by their luminescence.

taken up in detail in Chapter 4. It should be pointed out that prints photographed with proper choice of filters show very frequently much better ridge detail than that one sees on inspection. Photographic exposures needed to develop prints are arrived at, in most cases, by trial and error. If the background luminescence differs substantially in color from that of the latent print, and it often does, then color photography can be fruitful. Color contrast can be achieved by judicious choice of treatment of latent prints with luminescent materials or chemicals which react with fingerprint residue to form luminescent reaction products. These topics will be addressed in subsequent sections of this chapter and in Chapter 4. It is assumed that forensic analysts are well versed in photography. This topic will, therefore, not be treated further. It should be pointed out at this juncture that some of the color plates (prints made from color slides) in this text show somewhat distorted colors (color difference of Figures 3.2 and 3.4 in comparison with Figure 3.3). This distortion is a function of exposure, the characteristics of the film, the unconventional nature of the light source and filters, and, most importantly, background luminescence. The color photographs shown in the text were taken with Ektachrome ASA 200 slide film. The yellow or orange background in some of the photographs arises from luminescence of the filter in front of the camera, resulting from incidence of scattered laser light on the filter [e.g., Figures 3.13 and 4.9 (Plates 6 and 7, pages 58 and 59)]. It should also be realized that some of the black-and-white and many of the color plates shown in this chapter have necessarily suffered considerable loss of detail during the course of reproduction from the original slides (Figures 3.4, 3.6, 3.9 to 3.12, and particularly 3.13).

Experience in the author's laboratory has shown that 3 W (all lines) laser power is a bare minimum for generally effective fingerprint detection. For research purposes, this power is adequate, but for use by forensic laboratories concerned with examination of large numbers of exhibits, lasers with 10 to 20 W power (blue-green, all lines) are more appropriate. Moreover, as will be described later in this chapter, it is worthwhile to have a UV option-equipped Ar-laser to obtain lasing in the 334 to 364 nm spectral region. To obtain lasing at these wavelengths with adequate power, a large Ar-laser (15 to 20 W, all lines visible) is needed.

Some aspects of the operation and maintenance of Ar-lasers are discussed in Chapter 5.

Several questions can be raised about the detection of latent prints by the above laser procedure:

1. Is the luminescence inherent to the fingerprint residue or is it due to contamination?
2. What surfaces are amenable to laser examination?
3. Can the laser method reveal latent prints not otherwise detectable?
4. How does the sensitivitiy of the method compare with conventional treatments?
5. Is there a fingerprint age limitation?

These questions are dealt with in the remainder of this section.

While fingers contaminated with luminescent material can give rise to laser-detected luminescence, a situation certainly not undesirable, it is not necessary to have finger contamination. Latent prints display inherent luminescence. For research purposes, one can ensure in several ways that a deposited fingerprint is not contaminated.

1. One expects that examination of hands under the laser prior to fingerprint deposition will reveal contamination by luminescers as luminescent smudges.
2. One can carefully wash hands (and then let some time elapse to allow for buildup of fingerprint material on the skin).
3. One can spectroscopically compare possible contaminants present in the laboratory with fingerprint residue.
4. One can examine fingerprints deposited by large numbers of individuals from various walks of life in several georgraphic locations.

Each of these procedures has been employed by the author and the results indicate that inherent fingerprint luminescence is usually observed during the course of examination of latent prints by the procedure of this section.

Fingerprints deposited on a variety of substrates, such as paper, plastics, glass, metals, etc., can be detected. Figures 3.2 to 3.4 (Plates 2 and 3, pages 54 and 55), show examples of laser-detected

Figure 1.14 Ar-laser/dye laser system in operation. (Courtesy of Spectra-Physics, Mountain View, Calif.)

Plate 2   54

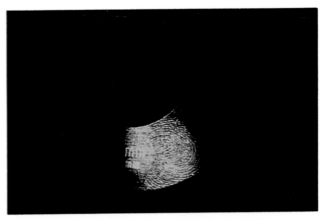

Figure 3.2   Laser-detected latent print on metal.

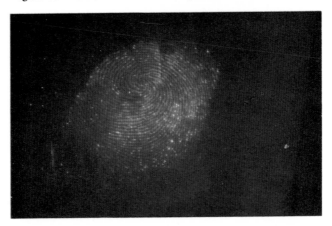

Figure 3.3   Laser-detected latent print on Styrofoam.

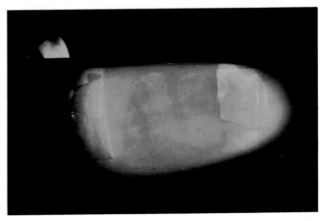

Figure 3.5   Laser-detected latent prints on living skin (see text).

Plate 3

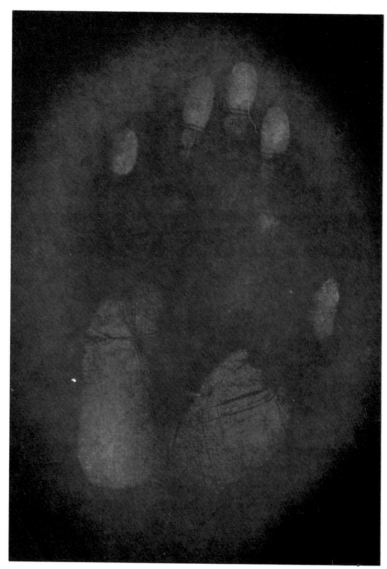

Figure 3.4   Laser-detected handprint on paper.

Plate 4

56

Figure 3.6 Latent print on black plastic dusted with Sirchie DP 002 powder and developed by laser.

Figure 3.7 Palm print on paper dusted with Mars Red and developed by laser.

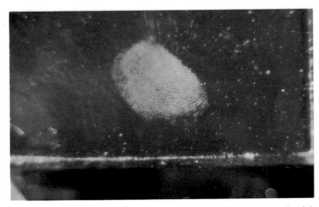

Figure 3.8 Latent print on transparent plastic dusted with Sirchie FMP 01 magnetic powder blended with rhodamine 6G and developed by Ar-laser.

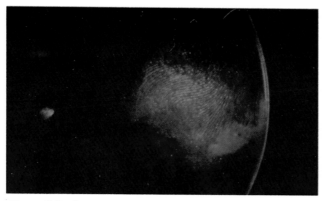

Figure 3.9 Latent print on glass dusted with FMP 01 powder blended with nile blue A perchlorate and developed by CW dye laser (6100Å).

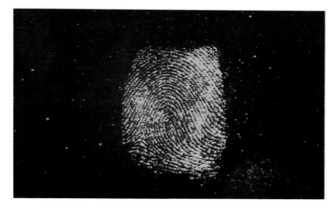

Figure 3.10 Latent print on black plastic stained with coumarin 6 and developed by Ar-laser.

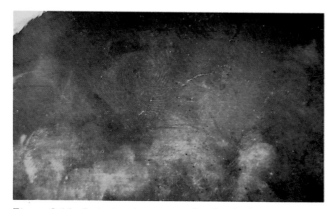

Figure 3.11 Latent print on metal stained with rhodamine 6G and developed by laser.

Plate 6   58

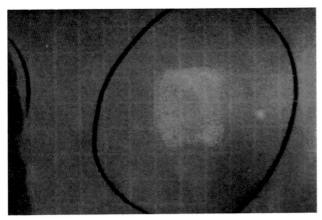

Figure 3.12 Latent print on paper treated with flouresca-mine and developed by laser.

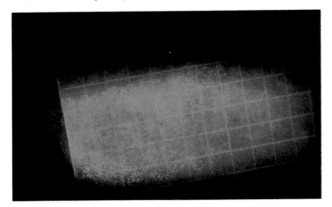

Figure 3.13 Latent print on paper treated with *ortho*-phthalaldehyde and developed by laser. Color distortion due to background emission and filter luminescence occurs.

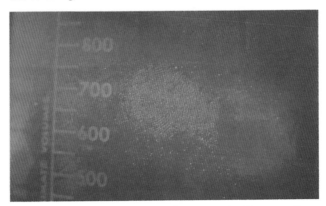

Figure 3.15 Latent print on glass treated with *para*-dime-thylaminocinnamaldehyde and developed by laser.

Figure 4.9 The strongest luminescent thin-layer chromatography bands of fingerprint material as developed by laser illumination. Because of background and filter luminescence, the band colors are distorted. Actual band colors are green, orange, and yellow from top to bottom.

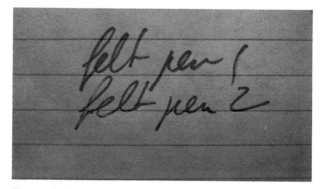

Figure 4.13 Writing from two felt pens as seen in room light.

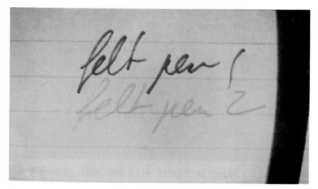

Figure 4.14 Writing of Figure 4.13 under laser light.

Plate 8

60

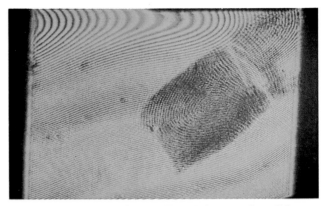

Figure 4.15 Ar-laser light reflected from a glass slide containing a latent print and projected onto a screen. Note interference patterns.

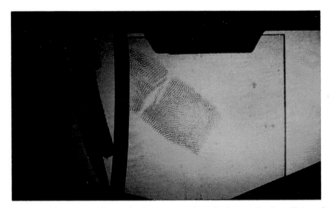

Figure 4.16 Ar-laser light transmitted through the slide of Figure 4.15 and projected onto a screen.

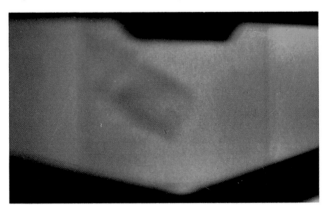

Figure 4.17 Same as Figure 4.16 but with ordinary light replacing laser illumination.

latent fingerprints on metal, Styrofoam, and paper, respectively. Some surfaces one would normally not consider suitable, such as some cloths, coarse paper towel, and living skin, have been found to be occasionally amenable to the laser method [5]. Figure 3.5 (Plate 2, page 54) shows the impression of three fingers on living skin (the forearm of a female). The photographic resolution shown in this figure is too low to reveal ridge detail which was, however, readily seen on closer observation of the impressions. The low-resolution photograph was chosen for presentation to provide an idea of the color of fingerprint luminescence as well as a feeling for the kind of contrast one frequently observes. An example of ridge detail of a print on living skin has been reported [5]. It should be noted that fingerprints on living skin tend to smudge out a short time after deposition because of skin perspiration. Latent prints are often considerably weaker than those shown in the above figures. They are nonetheless photographable if proper selection of filters is made.

To test the merit of the laser detection procedure in cases in which one does not have fresh latent prints, a print was placed on paper which was then baked at\75° for 2 weeks. The paper was subsequently soaked under running water for several minutes and left to dry. The latent fingerprint was still detectable by laser, but not by silver nitrate or ninhydrin [5]. While this result indicates that the laser method can reveal latent prints not otherwise detectable, it is not to imply that the laser procedure is necessarily always more sensitive than conventional methods. In fact, one frequently finds that latent prints not detected by laser are developed by ninhydrin. The converse is often found as well, however. The two methods, rather than competing, thus complement each other. Exhibits examined by laser are still amenable to all conventional treatments. If both laser inspection and conventional developments are to be used on an exhibit, it is recommended that the laser examination be performed first, since dusting, ninhydrin, and silver nitrate can have deleterious effect on inherent fingerprint luminescence.

Old fingerprints can be detected by laser. Prints as old as 10 years have been detected by the author. Results to date suggest that the color of fingerprint luminescence changes with fingerprint age, which brings to mind potential of the laser method in fingerprint age determination [5, 6]. This matter will be taken up in more detail in Chapter 4. It should be noted that prolonged laser exposure of latent fingerprints causes photodecomposition and concomitant fingerprint luminescence intensity decrease.

3.3 LASER DETECTION OF LATENT PRINTS—TREATMENT WITH FLUORESCERS

Even though the detection procedure of Section 3.2 has been found to have a wide range of utility, it fails in a number of frequently occurring situations because of one or more of the following reasons:

1. Strong substrate luminescence overwhelms the weak inherent fingerprint luminescence.
2. The substrate does not accept the fingerprint well.
3. Fingerprints deposited by a minority of individuals do not show sufficiently strong inherent luminescence. This situation is not so much a matter of the given individual, but a matter of his or her nervous state. Fingerprints deposited by excited or nervous individuals tend to luminesce better than they do when these individuals are calm. Thus, pronounced day-to-day variation in luminescence of fingerprints deposited by a given individual can be found.

A number of adaptations of the laser method to cases falling into these categories can be made [7]. One of these involves treatment of the latent print with fluorescent material followed by laser examination. The most obvious treatment in this category is dusting with fluorescent powders. Indeed, luminescent dusting powders to be used in conjunction with UV lamps, for instance Sirchie DP 002 and FMP 01, are on the market. Such powders would, offhand, not appear useful in conjunction with Ar-lasers (blue-green light). However, reasonably powerful Ar-lasers equipped with a UV option provide very substantial UV lasing (1 to 3 W) in the 334 to 364 nm range, so that excellent detectability of these powders when used with thus equipped lasers is obtained. The

Sirchie DP 002 white dual-purpose powder, though not designed for this, responds (weakly) to blue-green Ar-laser light. The observed luminescence color is green. However, when a lot of powder adheres to a latent print, the observed color of luminescence may appear orange. A print treated with Sirchie DP 002 powder and developed with blue-green Ar-laser light is shown in Figure 3.6 (Plate 4, page 56).

If a substrate shows substantial green luminescence, then the above powders do not provide good luminescence color contrast. A commercial dusting powder not designed for fluorescence, but which shows strong red luminescence under blue-green Ar-laser light, namely Mars Red (Criminal Research Products) can in such cases be used effectively. Combination of use of this powder with laser examination significantly enhances detectability over that obtained when dusted exhibits are simply inspected in room light [7]. Latent prints not developed by dusting with Mars Red and inspection in room light can be brought out under the laser. Figure 3.7 (Plate 4, page 56) shows a thus developed print. A highly valuable feature of dusting powders which display red luminescence when illuminated with blue-green Ar-laser light is that the luminescence color differs substantially from that of the substrate luminescence (yellow or orange) in the large majority of situations. This allows use of red-transmitting long-wavelength-pass filters which suppress yellow and orange, thus greatly enhancing contrast. Several powders which possess the necessary strong red luminescence to be useful as dusting materials have been reported [7].

Materials which have good luminescence properties, but which do not possess the necessary adhesion quality to form good dusting powders, can be blended with material which adheres well to fingerprint residue. Thus, a variety of luminescent dusting powders can be prepared. A blended powder, utilizing coumarin 6, a strongly fluorescent compound, has been reported by Thornton [8]. The following recipe has been successfully used by the author for a variety of dyes [9].

    1. Dissolve about 0.1 g dye in approximately 50 ml methanol.

2. Add roughly 10 g FMP 01 powder and stir until the methanol has evaporated.
3. Dry the residue for a short period of time in a warm oven.

Acridine yellow, rhodamine 6G, rhodamine B, nile blue A perchlorate, 3,3'-diethyloxadicarbocyanine iodide (DODC), 3,3'-diethylthiatricarbocyanine iodide (DTTC), and oxazine perchlorate have been blended with FMP 01 powder. This powder is a magnetic dusting powder, which, by itself, already shows green luminescence. Magnetic brush application consumes very little material, a convenient feature when one blends with dyes which are expensive. The latter four of the above cited dyes require red light for illumination if fluorescence is to be obtained. He-Ne lasers with about 50 mW power or CW dye lasers (pumped by Ar-lasers) can provide the red light. Alternatively, rhodamine 6G or B can be dissolved together with one of these dyes. The rhodamine, which responds to Ar-laser light, sensitizes the dye via energy transfer or absorption of rhodamine fluorescence by the dye followed by re-emission as dye fluorescence. Figures 3.8 and 3.9 (Plates 4 and 5, pages 56 and 57) show laser-developed latent prints dusted with FMP 01/rhodamine 6G and FMP 01/nile blue, respectively. A 6100 Å dye laser excitation was used for the latter. In view of recent reports of successful development of latent prints on skin with magnetic powder [10,11], luminescent magnetic powders take on particular potential value in examination of assault, rape, and murder victims.

In addition to dusting with fluorescent powders, staining with fluorescent dyes, when used in conjunction with laser examination, can develop latent prints. Laser dyes, by virtue of their intense fluorescence and frequent amenability to blue-green excitation, are obvious choices. In the reported applications, the exhibit was first dusted with finely ground coumarin 6 powder, which revealed no ridge detail on inspection in room light or under the laser. The exhibit was then sprayed with methanol to dissolve dye, and left to dry. A second, sometimes fairly vigorous, spraying was performed to remove excess dye from the substrate. The coumarin showed a preferential adhesion to the fingerprint residue, even though not to the extent that latent prints could be seen on inspection in room light. Under blue-green Ar-laser illumination, however, fingerprints were developed [5]. Since the second methanol spraying may have to be vigorous, the dye-staining method works better for older prints than for fresh ones. Fresh latents have a tendency to be washed off by strong methanol spraying. Rhodamine 6G and rhodamine B, which are also laser dyes, can be employed similarly. Figure 3.10 (Plate 5, page 57) shows a fingerprint on black plastic developed by coumarin 6 and Figure 3.11 (Plate 5, page 57) shows a fingerprint on metal developed by rhodamine 6G. Surfaces to which dye adhesion is very good are not amenable to the dye-staining procedure.

3.4   LASER DETECTION OF LATENT PRINTS—CHEMICALS REACTING WITH FINGERPRINT MATERIAL TO FORM LUMINESCENT REACTION PRODUCTS

Ninhydrin reacts with amino acids of fingerprint residue, forming a purple reaction product via which fingerprints are rendered visible. Thus stained latent prints do not show significant luminescence under UV or blue-green Ar-laser illumination. It would be useful to have chemicals such that similar reactions would lead to luminescent products amenable to Ar-laser excitation.

Two such chemicals, fluorescamine, first reported by Udenfriend et al [12] and *ortho*-phthalaldehyde, first reported by Roth [13], react with primary amines, i.e., amino acids, peptides, and proteins, to form reaction products which luminesce on illumination with UV light. These chemicals have been applied to the detection of latent fingerprints using UV lamps for exhibit illumination [14]. Recipes reported by Lee [14] are:

Fluorescamine

1. Dissolve 20 mg fluorescamine and 0.4 ml triethylamine in 100 ml acetone.
2. Spray exhibit and illuminate with UV light.

Ortho-Phthalaldehyde

1. Dissolve 2.5 g boric acid in 95 ml distilled water.
2. Adjust pH to 10.4 with 6 M KOH.

3. Add 0.3 ml 30 percent Brij 35 detergent.
4. Add 0.2 ml 2-mercaptoethanol.
5. Dissolve 240 mg *ortho*-phthalaldehyde in 2 ml methanol.
6. Add step 5 to the solution made by steps 1 to 4; mix well.
7. Spray exhibit, then illuminate with UV light.

Fluorescamine and *ortho*-phthalaldehyde, which are commercially available under the trade names Fluram (Roche Diagnostics) and Fluoropa (Durrum), respectively, have also been successfully used with a UV option-equipped Ar-laser [7]. Because of the high UV powers achievable with such lasers, detectability with the above treatments can be much improved over that obtained with UV lamps. A fingerprint on paper developed by Fluram and laser examination is shown in Figure 3.12 (Plate 6, page 58). A palm print on paper, developed by *ortho*-phthalaldehyde and laser examination, is shown in Figure 3.13 (Plate 6, page 58).

A further chemical which reacts in a manner similar to the above two compounds and ninhydrin is *para*-dimethylaminocinnamaldehyde, reported by Morris [15]. This material, much like ninhydrin, gives a reddish purple stain when reacting with the urea of fingerprint residue. In addition, the reaction of the compound with fingerprint material leads to orange latent print luminescence under blue-green Ar-laser excitation [7]. The author has used the recipe [16]:

1. Dissolve 1.5 g toluenesulfonic acid in 100 ml acetone.
2. Dissolve 0.5 g *para*-dimethylaminocinnamaldehyde in 100 ml acetone.
3. Mis solutions 1 and 2 and spray exhibit.

A sketch of the spray bottle (Canlab) used by the author is shown in Figure 3.14.

Even when latent prints thus treated do not exhibit discernible reddish purple staining in room light, ridge detail may still be quite apparent on laser examination [7]. A print in this category is shown in Figure 3.15 (Plate 6, page 58). A variety of surfaces (paper, cardboard, wood, glass, etc.) are amenable to the treatment. A convenient feature of the method is that the normal mode of Ar-lasing can be employed. However, if substrates show

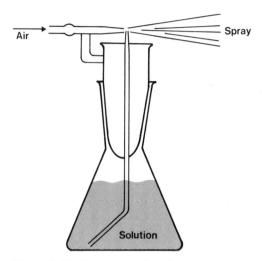

Figure 3.14 Pyrex spray bottle.

strong orange luminescence, then the green luminescence obtained with fluorescamine, for instance, may be more effective. The use of *para*-dimethylaminocinnamaldehyde together with a variety of acids has been reported by Sasson and Almog [17]. As in the detection of latent prints by inherent luminescence or treatments with fluorescent materials, the success of the procedures of this section may require careful choice of filters for inspection and photography. Filters will be taken up in detail in the next chapter.

As with ninhydrin, time between treatment and ridge detail development may be required when chemical procedures are used and heating may expedite development. As reported, some dusting materials may be somewhat poisonous [7]. Chemicals used for latent print development could possibly have carcinogenic properties. Thus, a few simple precautions, such as the use of gloves and confinement of materials to fume hoods during latent print treatment and storage, are recommended. It should be recognized as well that some chemicals may lose their effectiveness on prolonged exposure to moisture, oxygen, or light. Therefore, chemicals should be stored in opaque containers, in dessicators, and, when necessary, under inert atmospheres such as nitrogen.

The substances which give rise to inherent fingerprint luminescence and the luminescent reaction products of this section tend to photodecompose on prolonged laser illumination. Thus, laser irradiation of samples under scrutiny should be kept to a minimum prior to photography.

3.5 LASER DETECTION OF LATENT PRINTS— TREATMENT WITH PHOSPHORESCERS

In the procedures described in Sections 3.2 to 3.4, fluorescence from the substrate holding the latent print can be problematic. For instance, fluorescamine may fail if green or yellow substrate emission occurs. The same applies to inherent fingerprint luminescence. *para*-Dimethylaminocinnamaldehyde can fail if orange substrate fluorescence is present. Surfaces in this categoy are certain papers, cardboards, woods, plastics, etc. It would be useful to have a laser detection procedure which allows total elimination of the substrate luminescence, which is generally fluorescence. A latent print treatment leading to *phosphorescence* would allow this. Such a procedure has been reported [18]. In terms of the luminescence lifetime (see Chapter 2), we will distinguish between fluorescence and phosphorescence on an operational basis: We call the luminescence fluorescence if the lifetime is shorter than 10^{-6} sec. We call the luminescence phosphorescence if the lifetime is 10^{-5} sec or longer, preferably 10^{-3} sec or longer. We do not consider delayed molecular fluorescence in detail in this text. Delayed fluorescence can arise from thermal excitation of molecules from the lowest triplet state (see Chapter 1) back to the first excited singlet state, whereby fluorescence with a lifetime corresponding to the triplet lifetime (the lifetime of phosphorescence) is generated. Alternatively, pairwise interaction between molecules in the lowest triplet state can result in delayed fluorescence which has a lifetime corresponding to half the triplet lifetime. Finally, delayed fluorescence can arise through energy transfer from a long-lived species to a fluorescer in systems containing several compounds. For our purposes, delayed fluorescence is dealt with as if it were phosphorescence.

If a latent print treatment yielding phosphorescence is available, then laser detection can be performed by the procedure shown in Figure 3.16. Here, the laser beam, suitably expanded with a lens, illuminates the treated exhibit. The exhibit sits inside a cylindrical light chopper. Such a chopper consists simply of a rotating hollow cylinder with slots as shown in Figure 3.16 (top view). The cylinder rotates in the direction shown by the arrow in the figure and periodically interrupts the incident laser light with a frequency which depends on the speed of rotation and the number of slots in the cylinder. During the time the laser light illuminates the exhibit, the camera (or eye) does not see the exhibit. The exhibit comes into view at a time after light cutoff. By this time, the substrate fluorescence has already decayed completely so that only the latent print phosphorescence is seen.

The Sirchie DP 002 powder mentioned earlier, a white powder, is described by the manufacturer as highly fluorescent when exposed to UV radiation. A yellowish green luminescence is indeed observed. However, the luminescence has a long-lived component. Once the incident UV light is shut off, a distinct afterglow is very clearly observable. Within the framework of the present section, this powder acts thus as a phosphorescent one. Sirchie Fluoromag latent print powder (FMP 01), a gray magnetic

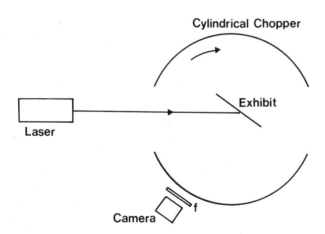

Figure 3.16 Apparatus for laser detection of latent prints by phosphorescence. f = filter (see text).

dusting powder, behaves identically in terms of its luminescence to the DP 002 powder. The green luminescence of the magnetic powder was examined for lifetime using a Princeton Applied Research boxcar averager (see Chapter 2). The afterglow spectrum was found to be the same as that of the prompt fluorescence (undelayed), which means that one has delayed fluorescence, rather than phosphorescence. The afterglow decay was found to be nonexponential, resolvable into two exponential components of lifetime 0.9 and 13 msec, with the longer-lived component corresponding to 30 percent of the afterglow in intensity. The afterglow intensity was found to be about 20 percent of the total luminescence intensity.

The detectability of latent prints treated with the FMP 01 or DP 002 powders can be enhanced when these powders are used with the procedure of Figure 3.16. To assess the sensitivity of this procedure, a fingerprint on yellow paper was dusted with DP 002. Fingerprint ridge detail was not observable in room light. On exposure to blue-green light from an Ar-laser and inspection as described in Sections 3.2 to 3.4, no ridge detail could be observed either because of the very strong yellow fluorescence of the paper. In fact, yellow paper was deliberately chosen as the test surface because of its strong fluorescence; blue-green light, to which the powder responds not nearly as well as to UV light, was chosen deliberately as well. These choices were designed to mimic a highly unfavorable situation. The sample was placed into a cylindrical light chopper (Princeton Applied Research Model 125 A, with the normal chopper wheel replaced by a homemade cylinder with two slots) and illuminated with the visible Ar-laser light. The latent print emerged clearly by its green afterglow; all substrate fluorescence had indeed been eliminated. Figure 3.17 shows a photograph of this print. Even if latent print phoshorescence or delayed fluorescence is extremely faint, so faint that the print needs to be located spectroscopically (as described in Chapter 4), photography presents no problem since arbitrarily long exposures can be used. The laser safety filter in Figure 3.16 is in principle not required, and only serves to block laser light which might spuriously be scattered around the room.

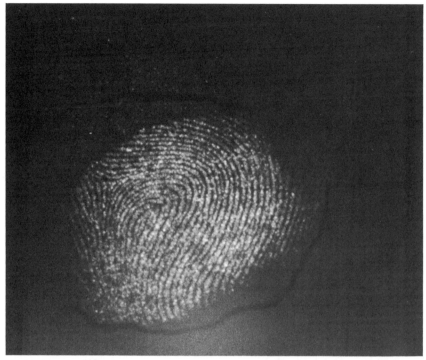

Figure 3.17 Latent print on strongly fluorescent paper dusted with DP 002 powder and developed in the phosphorescence mode of Figure 3.16

Hopefully, chemical treatments will eventually be developed, such that the phosphorescence detection procedure can be applied in instances in which the above powders fail.

3.6 CASE APPLICATIONS

Successful case applications of laser detection of latent prints are beginning to emerge from forensic laboratories, such as those of the Ontario Provincial Police and the FBI [19]. In this section, three case applications carried out in the author's laboratory are described. Each of these cases has a unique feature which demonstrates the power of the utilized procedure.

Inherent Luminescence

The first successful case application ever achieved by the laser method involved a piece of black electrical tape connected with a drug investigation conducted by the Ontario Provincial Police. The exhibit was brought to the author's laboratory for examination by laser. It was suspected that the tape might contain a latent print on its sticky side. An interesting feature of this exhibit was its unsuitable nature in terms of conventional methods. Obviously, it could not be dusted. The dark color of the surface would not give adequate contrast with ninhydrin or silver nitrate. In addition, the involved type of tape tends to decompose when exposed to a number of solvents and it deforms on stress or heating. The tape was exposed to blue-green Ar-laser light and inspected through laser safety filters. Care was taken to keep the light flux sufficiently low not to warp the tape by overheating. A very faint greenish yellow hue, the usual color of inherent fingerprint luminescence, was observed, but not adequate ridge detail was apparent. The exhibit was nonetheless photographed through a laser safety filter. Because of the obvious weakness of luminescence, long exposures (of some 20 min) had to be made. A print with ridge detail sufficient for identification and corresponding to a fingerprint of a suspect in the case was developed. The print is shown in Figure 3.18. A major feature of this case was the demonstration that the laser method can be more sensitive than the eye, i.e., fingerprints too weak to be observable by exhibit inspection may be photographable. While one would think that fingerprints too weak to be seen by inspection are worthless from a practical standpoint, Chapter 4 will treat a method which can allow detection of such prints. The exhibit further demonstrates an intrinsic advantage of luminescence photography over photography in the conventional fingerprint development sense. In conventional photography, both the substrate and the latent print scatter light which reaches the camera. The light scattered from the substrate limits the exposure time one can apply, hence the sensitivity. In luminescence photography under reasonably ideal conditions (low substrate luminescence), which can often be achieved through judicious choice of filters, long exposures allow

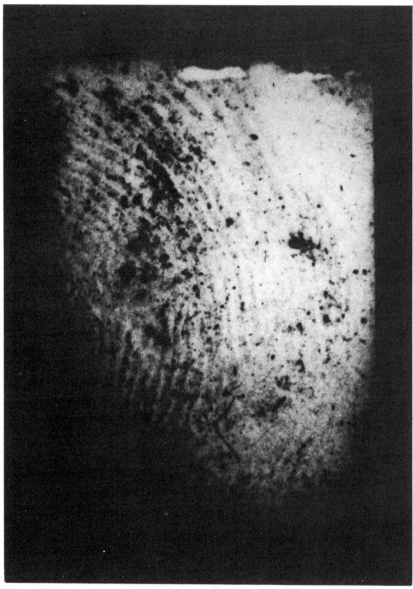

Figure 3.18 Latent print on sticky side of black plastic electrical tape; detected by inherent latent print luminescence.

high sensitivity to be obtained because only light from the latent print itself reaches the camera. In a similar way, luminescence spectroscopy is intrinsically more sensitive than absorption or reflectance spectroscopy.

One should note that prints made from black—and—white negative film show ridge detail as white on a dark background, the opposite of the conventional situation (see Figure 3.17). One might consider the possibility of treating latent prints on a strongly luminescent background with materials which strongly absorb laser light, but which do not luminesce. This would yield dark ridge detail on a light background. While such procedures may occasionally be fruitful (and should be explored), the situation here reflects the conventional situation, and is generally not expected to provide the high sensitivity one obtains when observing luminescent latent prints.

Mars Red

A .22 caliber rifle, the suspected murder weapon in an investigation conducted by the Royal Canadian Mounted Police in British Columbia, was examined by laser. No prints could be detected by inherent luminescence. The poor condition of the rifle (dirty and rusty) did not aid the examination. As varnished woods often do, the gun stock exhibited a very strong orange luminescence under laser light (visible) which overwhelmed any possible weakly luminescent latent print. The stock was therefore dusted with Mars Red and inspected in room light. Still, no fingerprints were discernible. The stock was then subjected to laser examination. A red—transmitting filter which blocks orange light was used for inspection. A print with good ridge detail was observed and photographed [7]. Though the ridge detail appeared sufficient to make the print identifiable, the author has not been informed of the disposition of the print. A photograph of the print is shown in Figure 3.19. A major point demonstrated by this exhibit is that dusting combined with laser examination can lead to greatly enhanced sensitivity. The case further demonstrates that laser detection can be effectively adapted to situations not conducive to inherent fingerprint luminescence detection.

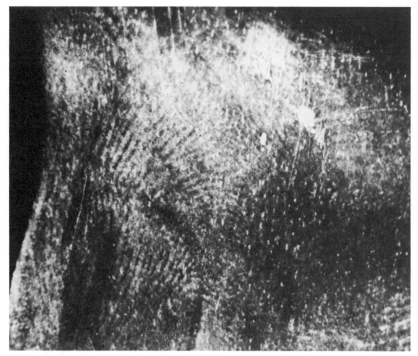

Figure 3.19   Latent print on gun stock; dusted with Mars Red and developed by laser.

para–Dimethylaminocinnamaldehyde

A cigarette pack, an exhibit involved in a murder investigation conducted by the Ontario Provincial Police, was unsuccessfully examined for inherent luminescence. It was then dusted with Mars Red which, too, failed to reveal any latent print. The exhibit was then treated with *para*–dimethylaminocinnamaldehyde. Very poor ridge detail was found on inspection of the exhibit in room light. The exhibit was subjected to laser examination. A print with good ridge detail (corresponding to the very poor detail observed in room light) was observed and photographed. This print, shown in Figure 3.20, was identified as, unfortunately, belonging to the victim. This case, again, demonstrates the increased sensitivity achievable by laser examination, and that exhibits can be examined sequentially with several procedures.

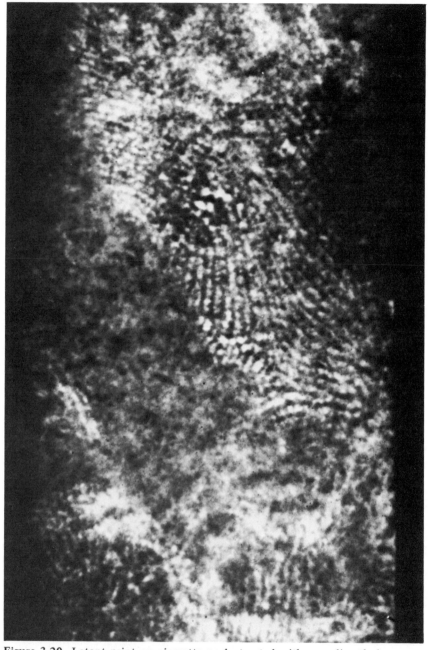

Figure 3.20 Latent print on ciagrette pack; treated with *para*-dimethyla-minocinnamaldehyde and developed by laser.

76

The chemical treatments, dusting procedures, and dye staining methods described in this chapter permit successful laser detection of latent prints on surfaces which luminesce, often strongly, and which are not generally well suited to detection by inherent fingerprint luminescence, and, in a number of instances, by conventional methods. Surfaces in this category are some woods, cardboards, and plastics, for instance. Dark surfaces, which often defy conventional procedures, are frequently amenable to laser procedures. There are, however, several disadvantages to laser methods. At this time, Ar-lasers with the required power are not portable, so that no field work can be done. Lasers are quite expensive. When optical spectroscopic techniques are used in conjunction with lasers (see Chapter 4) further technical complexity and cost result. Nonetheless, the value of lasers in detection of latent prints and other potential forensic applications (Chapter 4) should amply compensate for these drawbacks. Indeed, laser detection may in time supplant many of the present procedures, particularly if new treatments leading to phosphorescence, further chemical treatments, and applications of spectroscopic techniques are brought to bear on the subject.

To date, research of the treatments of this chapter has primarily aimed at demonstration of feasibility. Considerable work is still needed to optimize procedures for a variety of surfaces.

REFERENCES

1. R. D. Olson, *Fingerprint and Identification Magazine 53*(10), 3 (1972).

2. G. L. Thomas, *J. Physics,* E. : Scientific Instruments *11,* 722 (1978).

3. A. M. Knowles, *J. Physics,* E : Scientific Instruments *11,* 713 (1978).

4. G. G. Guilbault (Ed.), *Practical Fluorescence, Theory, Methods, and Techniques,* Marcel Dekker, New York, 1973, p. 334.

5. B. E. Dalrymple, J. M. Duff, and E. R. Menzel, *J. Forensic Sci. 22* (1), 106 (1977).

6. J. M. Duff and E. R. Menzel, *J. Forensic Sci.* 23(1), 129 (1978).

7. E. R. Menzel and J. M. Duff, *J. Forensic Sci.* 24(1), 96 (1979).

8. J. I. Thornton, *J. Forensic Sci.* 23(3), 536 (1978).

9. E. R. Menzel and K. E. Fox, *J. Forensic Sci.* 25(1), 150 (1980).

10. G. J. Reichardt, J. C. Carr and E. G. Stone, *J. Forensic Sci.* 23(1), 135 (1978).

11. R. F. Hall (Dade County Public Safety Dept., Miami, Florida), case report presented at the 1979 Semi-Annual Educational Seminar of the Florida Division of the International Association for Identification, Tallahassee, Florida, May 24 to 26, 1979.

12. S. Udenfriend, S. Stein, P. Bohlen, W. Dairman, W. Leimgruber, and M. Weigele, *Science 178*, 871 (1972).

13. M. Roth, *Anal. Chem. 43*, 880 (1971).

14. H. C. Lee (University of New Haven, Connecticut), paper presented at the 63rd Annual Educational Conference of the International Association of Identification, Austin, Texas, July 23 to 27, 1978.

15. J. R. Morris, British Patent 1423025, 1976.

16. C. W. Melton (Battelle Laboratories, Columbus, Ohio), personal communication.

17. Y. Sasson and J. Almog, *J. Forensic Sci.* 23(4), 852 (1978).

18. E. R. Menzel, *J. Forensic Sci.* 24(3), 582 (1979).

19. B. E. Dalrymple (Ontario Provincial Police), J. F. Walters (FBI), personal communication; B. E. Dalrymple, *J. Forensic Sci.* 24(3), 586 (1979).

4 APPLICATIONS OF SPECTROSCOPY TO LATENT FINGERPRINT DETECTION

This chapter describes several applications of spectroscopic measurements of the type presented in Chapter 2 to the study of the luminescence of latent fingerprint residue and the detection of latent prints. These measurements have two purposes, namely to provide insight into the nature of the latent print luminescence and to optimize detectability. Section 4.1 describes absorption, emission, and excitation spectra of fingerprint material. Based on spectroscopic results, optimized choices of filters for best photography of latent fingerprint luminescence (either inherent or arising from the treatments described in Chapter 3) are treated in Section 4.2. Section 4.3 considers the nature of luminescers in fingerprint residue. Section 4.4 presents experimental procedures designed to permit detection of latent fingerprints when their luminescence is too weak under the laser to be discerned by eye. Finally, Section 4.5 considers possible future applications of lasers and optical spectroscopy to forensic analysis on a wider scope.

4.1 LATENT FINGERPRINT LUMINESCENCE

Two questions arise immediately when one considers the method of Ar-laser examination of exhibits and the observation of inherent fingerprint luminescence through laser safety filters. These are:

1. Is one observing luminescence adequately, i.e., is one observing wavelengths near the peak of luminescence, or is one just seeing the "tail" of the emission?
2. Is the Ar-laser in fact the best laser source for the method on hand?

79

These questions have been addressed by study of absorption, emission, and excitation spectra of fingerprint material [1, 2]. Spectra have been obtained of material which was removed from fingers with a methanol-soaked swab and then dissolved in methanol, or of fingerprints deposited on Mylar or glass. Absorption spectra revealed, in addition to the ubiquitous UV absorbance common to most organic materials, absorption (albeit weak) in the 400 to 530 nm spectral range. Emission and excitation spectra of dissolved fingerprint matter as well as fingerprints deposited on surfaces revealed that indeed the Ar-laser's blue-green lines are highly suitable for excitation. In observing emission through laser safety filters, some luminescence, namely that occurring at 470 and 500 nm [1], is lost. This loss is not overly serious, however, and can, in any event, not be avoided because the laser light scattered from the exhibit *must* be filtered out not only for safety reasons, but also to enable one to observe the faint latent print luminescence. Figure 4.1 shows the excitation spectrum cor-

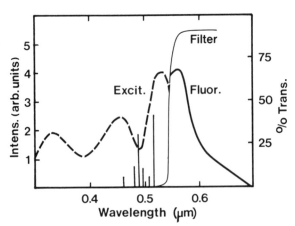

Figure 4.1 Fluorescence spectrum, prefiltered with a laser safety filter, and excitation spectrum of fingerprint on glass. The transmission characteristics of the laser safety filter are also shown. The vertical lines denote the Ar-laser lines, with line lengths representing relative intensities.

responding to the 550 nm emission (yellow-green) of fingerprint residue on glass. The Stokes shift (see Chapter 1) between absorption and emission is about 20 nm. It is seen that the laser wavelengths, which fall in the 450 to 520 nm range, with 488 and 514.5 nm the wavelengths of the strongest laser lines, are excellently suited for excitation. The fingerprint luminescence spectrum (as seen through laser safety filters) obtained with all-lines excitation is also shown in Figure 4.1. The transmission spectrum of the laser safety filter (Fisher 11-409-50A) and the principal laser lines are shown in this figure as well. The lengths of the lines denote the relative intensities of the Ar-laser lines. The data in this figure show that of the presently available CW lasers, the Ar-laser is the best choice for inherent fingerprint luminescence detection. The advantages of CW lasers over pulsed lasers have already been discussed.

4.2 FILTERS

The spectral features of latent fingerprint residue shown in Figure 4.1 indicate that the laser safety filter transmits unnecessary light in the red spectral region. Even orange transmission could be done without, particularly if orange substrate luminescence occurs, and it often does. A band-pass interference filter with peak transmission at about 550 nm should greatly aid maximization of contrast by eliminating background emission as much as possible. The characteristics of a band-pass filter are schematically shown in Figure 4.2. Interference filters, particularly those of relatively broad band width, transmit significantly outside their band-pass region. For example, interference filters of about 50 nm band width transmit on the order of 0.1 percent outside the band pass. This amount of light transmission is still very significant if it occurs at a laser wavelength, so that band-pass interference filters should be used together with long-wavelength-pass filters (such as laser safety filters) which block the laser light, as shown schematically in Figure 4.2. Interference filters with band width of about 10 nm are commercially available with peak transmittances obtainable in 10 nm intervals, so that the most suitable interference filter is easily chosen

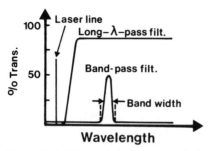

Figure 4.2 Transmission characteristics of a long-wavelength-pass filter serving as a laser safety filter and a band-pass interference filter. The vertical
broken lines represent the band pass. The band width denotes the wavelength
interval between points at which the filter transmits half the amount it transmits at the maximum-transmission wavelength.

once the fingerprint spectrum and, if necessary, the background
spectrum are on hand. The use of narrow-band interference filters is not always desirable for purposes of visual inspection because of the overall reduction of luminescence reaching the eye.
Transmission of interference filters is typically only 50 percent
at the peak of the band pass. For photography, however, where
exposure can always be suitably increased, the gain in achieved
contrast can be dramatic when such interference filters are employed. In many instances a band width of 50 to 100 nm suffices.
The choice of filters depends primarily on the wavelength of the
illuminating laser light (blue-green versus UV) and the features of
the latent print luminescence (inherent versus fluorescamine versus Mars Red, etc.) with respect to the color of the background
luminescence.

For dusting with Mars Red, for instance, a filter which only
passes wavelengths longer than about 600 nm is the choice to be
made [3].

For inherent fingerprint luminescence, a combination of laser
safety filter and interference filter with about 20 nm band pass
and peak transmission at about 550 nm should generally provide
the best contrast in photography.

Fluorescamine-treated fingerprints cannot be developed by

laser illumination with blue-green light, but can be brought out nicely with a UV option-equipped Ar-laser. The reason for this is demonstrated by the fluorescence and excitation spectra of a fluorescamine-treated primary amine, shown in Figure 4.3. The excitation spectrum shows that the normal visible Ar-laser lines cannot be absorbed by the fluorescent reaction product. However, UV Ar-laser light can be absorbed quite well. The fluorescence spectrum is broad and peaks at about 500 nm. Thus, a long-wavelength-pass filter transmitting wavelengths longer than about 400 nm, serving as a laser safety filter, together with a band-pass interference filter of 500 nm peak transmission and a band width of 50 to 100 nm should be used unless strong background luminescence dictates another choice.

The phosphorescent dusting powders (Sirchie DP 002 and FMP 01) discussed in Chapter 3 do not respond well to the blue-green light of Ar-lasers. The reason for the poor response is shown by the excitation spectrum for the FMP 01 powder, given in Figure 4.4. This spectrum shows that a UV option-equipped Ar-laser can excite this powder quite well. The DP 002 powder behaves identically to the FMP 01 powder in terms of luminescence. If the two powders are used with a UV option-equipped Ar-laser in the normal laser detection mode (Section 3.3), then a long-wavelength-pass filter transmitting at wavelengths longer than about 400 nm should be used as a safety filter. In addition, a band-pass interference filter of peak transmission in the 530 nm

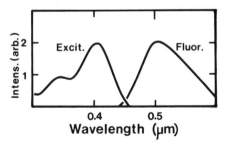

Figure 4.3 Excitation and fluorescence spectra of the fluorescent product of the reaction between fluorescamine and a primary amine.

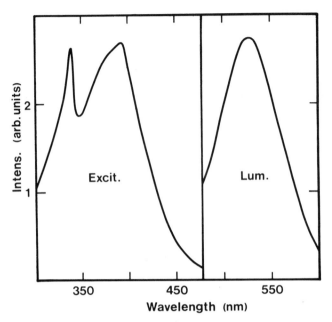

Figure 4.4   Excitation and luminescence spectra of Sirchie FMP 01 powder.

region (see Figure 4.4) and 50 to 100 nm band width should be helpful. If the powders are used in the phosphorescence detection mode (Section 3.5), then filters are not essential.

Figures 4.5 to 4.8 show transmission characteristics, measured with a Cary 17 spectrophotometer, for a number of useful filters when UV (3–72) or blue-green (3–67) laser light is used. The 2–63 filter is useful in conjunction with red-luminescing dusting powders. When using such filters together with high-power lasers and spectroscopic instrumentation, one should recognize that the filters themselves can luminesce when illuminated by laser light.

Figure 4.6 shows transmission curves for broad-band interference filters (Oriel). If such filters are on hand, but narrower band pass is desired, then two filters can be combined.

Examples of transmission curves of narrow-band-pass interference filters (Ditric Optics) are shown in Figure 4.7. The three
spectra, with peak transmission at 500, 550, and 650 nm, cor-
respond to filters useful for detection with fluorescamine or the
Sirchie powders (500 nm), inherent fingerprint luminescence (550
nm), and Mars Red (650 nm).

The wavelengths transmitted by an interference filter depend
on the orientation of the filter with respect to the incident light.
If an interference filter transmits at a given wavelength when the
filter surface is perpendicularly oriented to the incident light (the

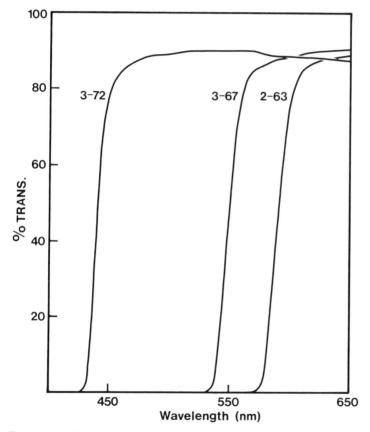

Figure 4.5 Transmission of Corning 3-72, 3-67, and 2-63 filters.

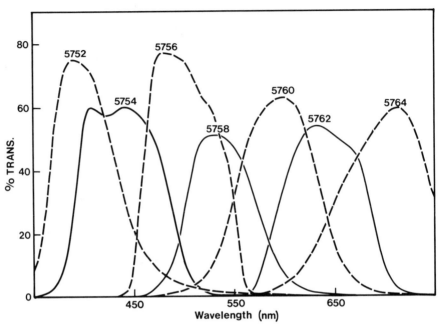

Figure 4.6 Transmission of Oriel 5752 to 5764 broad-band interference filters.

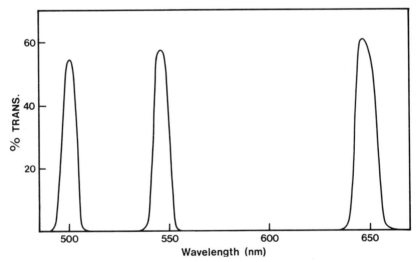

Figure 4.7 Transmission of Ditric Optics narrow-band-pass interference filters.

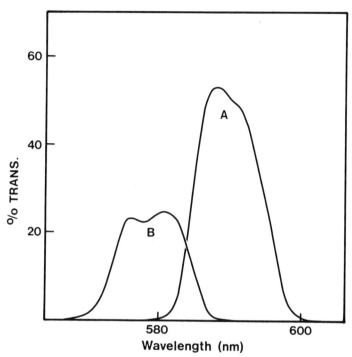

Figure 4.8 Transmission of an interference filter oriented perpendicular to the incident light (A) and tilted by about 30° (B).

usual orientation), the transmission can be shifted to shorter wavelengths if the filter is tilted. This tilting allows one to adjust the transmission characterisitcs of interference filters to some extent. Figure 4.8 shows the transmission change of a narrow-band-pass interference filter when it is tilted by about 30° (B) from the normal orientation (A).

4.3 NATURE OF LUMINESCERS IN FINGERPRINT MATERIAL

One might be curious about the nature of the species in fingerprint residue which cause the luminescence observed upon Ar-laser exhibit illumination. The known presence in fingerprint residue of riboflavin (a B-vitamin) [4] which fluoresces yellow-green [5] suggests that this compound might well be largely responsible for fingerprint luminescence. It is known that riboflavin is water soluble and tends to photodecompose. That fingerprint luminescence is often not seriously affected by moisture can be understood, nonetheless, since the riboflavin is likely buried in lipids and thus protected. Prolonged light illumination causes fingerprint luminescence degradation, as mentioned in Chapter 3, but not overly quickly, which again can be due to the protective environment of the riboflavin.

Excitation spectra for fingerprint material have been obtained [1]. They show a definite resemblance to the known excitation spectrum of riboflavin [6], but do not match precisely. Because of the profusion of chemicals in fingerprint residue, on the other hand, one can anticipate the possibility of substantial complexing so that a precise match need not necessarily occur.

Combination of thin-layer chromatography of fingerprint material with laser examination of developed chromatography plates has been reported [2]. Results show that fingerprint material contains a variety of luminescent species. Figure 4.9 (Plate 7, page 59) shows a photograph of the strongest luminescent chromatography bands as revealed under the laser. The remarkable similarity of the luminescence spectrum of the strongest band with the luminescence spectrum of a vitamin B tablet [2] further supports the identification of riboflavin as a likely major luminescer in fingerprint residue. Considerably more research by investigators versed in biological materials is necessary, however, to elucidate in detail the nature of luminescers in fingerprint residue.

In the reported work on thin-layer chromatography coupled with laser examination [2], developed chromatography plates showed *nothing* upon inspection in room light but a substantial number of luminescent bands under the laser. Analogous results were obtained with other exochrine fluids, such as urine and saliva. This suggests that chromatography together with laser spectroscopy may be of potential in forensic analysis.

4.4 DETECTION OF WEAK PRINTS

To test the sensitivity of spectroscopic detection of fingerprint luminescence, a very weakly luminescing print was deposited on glass. The print could not be observed at illumination levels less than 0.8 W/cm^2, an illumination level corresponding to 15 W of laser light illuminating a circular area of 2 in. diameter! The spectrum of this print could be readily measured spectroscopically, however, as will be described shortly, with 0.2 W/cm^2, i.e. a light level reduction by a factor 4, and instrument settings far from the ultimate detection limit. This detectability situation does not only apply to inherent fingerprint luminescence, but also to luminescence arising from latent print treatments. Since background subtraction, as described in Chapter 2, needs to be made, the background limits the spectroscopic sensitivity, which, nonetheless, is much superior to that of the eye. The above result brings to mind the possibility of locating faint (not discernible by eye) fingerprints spectroscopically, so that they may then be photographed. Recall that the first case application described in Chapter 3 clearly shows that photography can exceed the eye in sensitivity. Proper filter selection is critical, however. Indeed, the above-mentioned fingerprint on glass was first located spectroscopically and only subsequently observed visually when the illumination level was increased from 0.2 to 0.8 W/cm^2.

Three spectroscopic arrangements designed to locate such faint latent prints on exhibits are presented in the ensuing descriptions.

1. In the luminescence spectroscopic arrangement shown in Figure 2.10 (see Chapter 2), the monochromator is set at the wavelength of latent fingerprint luminescence (about 500 nm for fluorescamine, etc.). The exhibit is then vertically and horizontally translated (or rotated in case of a rounded exhibit) such that the position of the illuminated spot does not shift with respect to the monochromator. This means that the optical alignment is not changed during exhibit movement. At the location of the fingerprint the signal displayed by the photon counter should increase. Because of light-scattering effects, the background emission at the location of the latent fingerprint can be reduced in comparison to the background emission one has when a bare spot of the exhibit is illuminated. Reasons for this, and the proper background subtraction procedure for measurements of spectra, have been described in Section 2.6 of Chapter 2. For the present purposes, the

possible reduction in background emission at the location of the fingerprint means that this reduction can compensate for the increase due to presence of fingerprint material. The measurement needs, therefore, to be repeated with the monochromator set at a wavelength at which the fingerprint does not luminesce, or at which it luminesces only weakly in comparison to the maximum wavelength. The second wavelength setting should be at a longer wavelength to avoid possible absorption of background emission by the fingerprint residue. If the background emission reduction has compensated for the fingerprint luminescence, then this second measurement, with translation oft the exhibit, will reveal the fingerprint by reduction of the signal level displayed by the photon counter. In the search for the earlier-mentioned fingerprint on glass, luminescence detection used a nine-stage photomultiplier tube of S-20 spectral response and a photon counting system comprised of Princeton Applied Research Model 1120 amplifier/discriminator and Model 1112 photon counter/processor. Photomultiplier dark-count subtraction was made. At the location of the latent print a 45 percent increase in luminescence signal was obtained, compared to signals from bare portions of the substrate, with the monochromator set at 550 nm. With the monochromator set at 650 nm, a spectral region in which latent prints are expected to luminesce only very weakly [2], however, a 10 percent signal decrease at the location of the latent print was found. To determine the fingerprint luminescence spectral maximum, the signal intensity from bare substrate illumination was set equal to that from the location of the fingerprint, with the monochromator set at 620 nm. This was accomplished by adjusting the photon counter sampling time for measurement at one location with respect to the other. Spectral scans were then run at the two locations. These are shown in Figure 4.10. The fingerprint spectrum, obtained as the difference between the observed spectra at the two locations, is also shown in Figure 4.10. This spectrum neglects weak fingerprint luminescence in the red spectral range.

The procedure is a somewhat laborious one. Additionally, power fluctuations in the illuminating light and fluctuations in the

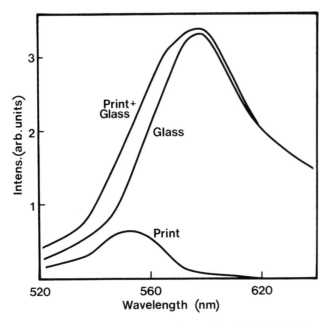

Figure 4.10 Luminescence spectrum of weak latent print on glass (see text).

optical alignment due to exhibit motion (primarily the latter) can be problematic. Instrumentation availability permitting, the procedure should be abandoned in favor of the following method.

2. An experimental arrangement which circumvents the above shortcomings is shown in Figure 4.11. This set up is employed as follows. One of the monochromators is set at the maximum fingerprint luminescence wavelength (550 nm for inherent luminescence, 500 nm for fluorescamine, etc.). The other monochromator is set at a wavelength which is considerably longer and at which little or no fingerprint luminescence occurs, but at which the substrate still emits. Should the fingerprint emit at wavelengths longer than those of substrate luminescence, an unlikely case, then a long-wavelength-pass filter can entirely dispose of background and the spectroscopic detection as well as photography of the fingerprint luminescence become rather simple. The signals from the two am-

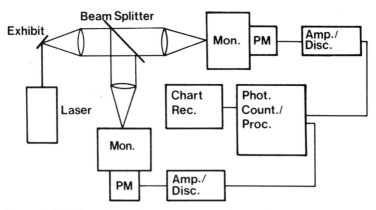

Figure 4.11 Experimental arrangement to locate latent prints spectroscopic-ally (see text).

plifier/discriminators (such as PAR Model 1120) of Figure 4.11 are fed to a two-channel photon counter/processor (such as PAR Model 1112) which displays the difference between counts from the two channels. The signal of one channel is *adjusted such that the displayed difference count is zero* when a bare location on the exhibit is illuminated. This null can be achieved in a variety of ways, such as introduction of variable neutral density filters (which attenuate light without distorting spectra), monochroma-tor slit width adjustment, etc. Once the null signal has been ob-tained, the exhibit is translated or rotated and the location of a latent print is revealed by a positive displayed signal which signi-ficantly exceeds the noise level.

Optical alignment problems are minimized by use of long focal length lenses. Since the beam splitter of Figure 4.11 is at 45° to the incident, parallel ray, light, the optical paths to the two mono-chromators are equivalent. Thus, small optical misalignment of the exhibit arising during its translation or rotation affects both mono-chromator inputs equally, which prevents optical alignment diffi-culties. The beam splitter should have approximately equal reflec-tance and transmittance. The two-channel mode of operation automatically compensates for excitation light intensity fluctua-

tion. Once a fingerprint is located, and, if necessary, the fingerprint luminescence spectrum is taken, carefully selected filters aid
the subsequent photography.

3. If a latent fingerprint treatment leading to phosphorescence
(see Section 3.5) is used, then the simpler of the above two experimental arrangements can be adapted. The photon counter is now
operated in the gated mode. The illuminating light is chopped with
a mechanical light chopper which does not need to be a cylindrical one. Counts are accumulated by the photon counter only during times after illumination cutoff. Because at these times the
fluorescence from the substrate has already decayed, no background subtraction is necessary. The experimental arrangement is
that shown in Figure 2.13.

While the content of this chapter thus far has been restricted
to description of spectroscopic measurements explicitly applicable
to latent fingerprint study and detection, there is no compelling
reason why laser spectroscopy should not be useful in forensic
work on a wider scope. Some applications in this category are suggested in the next section.

4.5 PROSPECTS FOR LASERS AND SPECTROSCOPY
IN FORENSIC ANALYSIS

The development of lasers was closely followed by a pronounced
increase in the vitality of spectroscopy in physics and chemistry,
such that by now the laser has taken an eminent place in these
fields. Lasers have become valuable also in medicine, machining,
communications, surveying, military applications, etc. There is
then little reason to doubt that lasers will likewise find a variety of
applications in forensic analysis.

A number of advances, which are more a matter of probability
than dream, can be anticipated. New chemical fingerprint treatments superior to present ones, leading to laser-detected luminescence, can be expected. With regard to chemical treatments, per-
usal of the biochemistry and analytical chemistry literature should
be valuable. Fruitful application of lasers other than Ar-lasers

should develop, particularly dye lasers and pulsed lasers operating
with high-power and fast pulse repetition rates. By combining
laser excitation and spectroscopic measurements, treatments
which lead to infrared emission, a case in which visual inspection
cannot be performed, may become useful.

When considering the application of lasers to fingerprint detection, one is invariably drawn to wonder about prospects for a
portable laser latent print examiners could take to crime scenes for
on-the-spot fingerprint search. Chapters 3 and 4 make it clear that
such lasers will, for the time being, likely have to be Ar-lasers.
Present-day Ar-lasers, unfortunately, require high electrical power
and water cooling, and are thus not portable. Hopefully, however,
portability will eventually be achieved. Some limited success may
be found with small Ar-lasers or He-Cd lasers, which are portable,
when laser illumination is combined with latent print treatments
akin to those described in Chapter 3. He-Ne lasers can be portable,
provided 110 or 220 V electrical power is available at the examination scene, and can deliver power of about 50 mW. Such lasers
can be used for latent print detection together with dusting powders which respond to red light. DODC, DTTC, oxazine perchlorate (see Chapter 3) are examples of dyes which are amenable to He-Ne laser excitation and which have been blended with
dusting powder [7]. Since these dyes fluoresce primarily in the
near-infrared, latent prints can generally not be located by simple
inspection, but require spectroscopic detection instrumentation.
Such instrumentation may be as simple as a photomultiplier (pre-
ceded by a suitable filter) and simple photon counting apparatus,
or can be more elaborate, as described in Section 4 of this chapter.
Infrared viewers may be of utility also. Combination of such a
detection system with a He-Ne laser which delivers about 50 mW
of power constitutes a possibly portable system. However, it does
not offer the sensitivity large Ar-lasers can provide.

Previous observation [1] suggests that inherent latent fingerprint luminescence changes color with age. The basic necessities to
study this spectral change with age, namely the ability to measure
fingerprint spectra and to isolate luminescers of fingerprint material, are accessible [2]. That fingerprint luminescence changes color

with age can be understood, at least partly, by the recognition that photochemistry occurs. For instance, prolonged light exposure leads to reduction of fingerprint luminescence strength. If riboflavin is indeed a major luminescer in fingerprint material, as strongly indicated by the data described earlier, then photochemistry can certainly be anticipated, sinc riboflavin is known to decompose on UV exposure. Thin-layer chromatographic bands of fingerprint material [2] undergo photochemistry, which is manifested by the disappearance of luminescent bands and the appearance of new luminescent bands when chromatography plates are exposed to UV irradiation and laser examination. If developed chromatography plates are examined by laser some months after development, one finds that all luminescence has vanished. This suggests that the ambient atmosphere, most likely through oxidation, can cause changes in fingerprint luminescence. Temperature, as well, can be expected to influence fingerprint aging. Figure 4.12 shows luminescence spectra (with excitation at 5145 Å) of a fresh latent print, a fresh latent print after exposure to 7 W/cm^2 of Ar-laser light (5145 Å) for some 15 min., and a latent print roughly a decade old from a page of a book. Because of the above-cited factors affecting fingerprint aging, latent print luminescence study as a function of age will be difficult. However, given the value of fingerprint age information to forensic analysis, such a study is well warranted, since at present no methods at all are available to determine the age of a latent fingerprint. In study of age-change of fingerprint luminescence, spectroscopy as described in this text will definitely be required. One should recognize that it may be that dietary factors play a role in fingerprint luminescence characteristics and that luminescence spectral distortion by contaminants can occur. If so, then fingerprint age determination may be quite difficult.

The sensitivity of spectroscopy, particularly when the unique capabilities of photon counting and lasers are taken advantage of, suggests potential analytical applications. Combination of lasers and spectroscopy with chromatography generally might prove useful. Perhaps an application to explosives tagging exists.

Many inks show strong luminescence under laser light. Indeed,

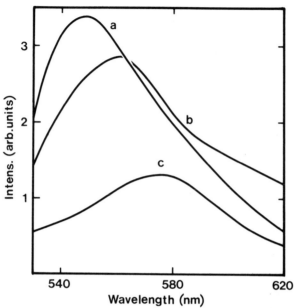

Figure 4.12 Luminescence spectrum of fresh latent print (a), fresh latent print after 15 min exposure to 7 W/cm^2 laser light (b), and latent print approximately 10 years old on paper (c). Laser illumination was at 5145 Å.

inks from some pens, which cannot easily be distinguished on inspection of writing in room light, can show dramatic differences on laser examination. Such a situation is shown in Figures 4.13 and 4.14 (Plate 7, page 59). The colors in these photographs are slightly distorted. Application of lasers and spectroscopy to document examination may prove fruitful.

The coherent nature of laser light gives it quite extraordinary properties, as evidenced by the field of holography (three-dimensional images). Scattering of laser light as well as transmission of laser light through transparent media can show some unique features when compared to similar illumination with ordinary lamps. Figures 4.15 and 4.16 (Plate 8, page 60) show Ar-laser light reflected by a glass slide containing a latent fingerprint and projected onto a screen, and laser light transmitted through this slide and projected onto a screen, respectively. The counterpart to the projection in transmission, when an ordinary lamp was used and the distance between slide and screen was about the same as that maintained during the laser illumination, is shown in Figure 4.17 (Plate 8, page 60). The superiority of laser light scattering is obvious. Interference patterns such as those seen in Figure 4.15 (Plate 8, page 60), are often observed when laser light illuminates transparent materials. Such patterns may find application in identification of materials. A nice study on reconstruction of fractured glass by laser-beam interferometry has been reported by Thornton and Cashman [8]. Laser light reflected from metallic surfaces can show very detailed patterns as well. Might lasers find an application in firearms examination?

An intriguing prospect is provided by a combination of laser, spectroscopic instrumentation, and an optical microscope. If laser light is focussed onto a small region of an exhibit under examination, the illuminated spot is in the focus of the microscope and the eyepiece is coupled to a spectrometer, then both spectral and structural information on a microscopic spatial scale becomes accessible. Luminescence or Raman spectra can provide chemical information, while ordinary scattering can yield structural detail. A somewhat more futuristic extension of this concept entails the idea of laser scanning of exhibits to obtain fingerprint ridge detail spectroscopically. Figure 4.18 schematically shows the concept. Laser light enters through the eyepiece of the microscope and is focussed on a small region of the exhibit. Fluorescence or Raman scattering is collected from this spot by the same microscope and exits through the eyepiece. The radiation is then incident on a spectroscopic system. If spectral information is desired, this includes a monochromator. Otherwise, a filter, photomultiplier, and photon-counting apparatus may suffice. The output of the spectroscopic system is analyzed by computer. The microscope scans the exhibit, thus scanning fingerprint ridges. Once computer analysis which, in addition to scan control, might entail image intensification, noise rejection, spectral analysis, etc., is

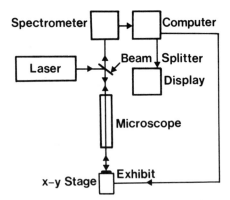

Figure 4.18 Proposed arrangement for scanning ridge detail (see text).

completed, the results, namely the latent print ridge detail and spectral information, are displayed on a cathode-ray tube or other display unit.

Undoubtedly many more applications can be foreseen. Perhaps the utility of lasers and optical spectroscopy in forensic work is primarily limited only by the ingenuity of forensic researchers and analysts.

REFERENCES

1. B. E. Dalrymple, J. M. Duff, and E. R. Menzel, *J. Forensic Sci.* 22(1), 106 (1977).

2. J. M. Duff and E. R. Menzel, *J. Forensic Sci.* 23(1), 129 (1978).

3. E. R. Menzel and J. M. Duff, *J. Forensic Sci.* 24(1), 96 (1979).

4. R. D. Olsen, *Fingerprint and Identification Magazine* 53(10), 3 (1972).

5. G. G. Guilbault (Ed.), *Practical Fluorescence, Theory, Methods and Techniques,* Marcel Dekker, New York, 1973, p. 334.

6. E. C. Lim (Ed.), *Molecular Luminescence,* Benjamin, New York, 1969, pp. 575-576.

7. E. R. Menzel and K. E. Fox, *J. Forensic Sci.,* 25(1), 150 (1980).

8. J. I. Thornton and P. J. Cashman, *J. Forensic Sci.* 24(1), 101 (1979).

5 OPERATION OF AR-LASERS

Details of operation, maintenance, and troubleshooting are presented in instruction manuals accompanying Ar-lasers. In this
chapter, we therefore only comment briefly on those aspects of
the operation of an Ar-laser which a user should be most familiar
with in order to maintain satisfactory instrument performance.
Troubleshooting, for which detailed procedures and circuit diagrams are provided in instruction manuals, will not be treated.
Such treatment requires extensive consideration of electronic circuitry, and is beyond the scope of the text. We note, though, that
it is worthwhile to have some spare parts, apart from line fuses and
laser supply fuses, on hand if the laser user has some expertise in
electronics. Spare parts might include power transistors (these are
large transistors mounted on a water-cooled heat sink in the laser
power supply and are probably the most frequently failing electrical elements), diodes for the three-phase full-bridge rectifier
circuitry, resistors, current regulator card, etc. Instruction manuals
sometimes list recommended spare parts. Otherwise the laser manufacturer can be consulted.

The installation, checkout, and, if necessary, alignment of a
newly purchased Ar-laser is generally handled by the vendor and
thus needs not be treated here. Three aspects of the initial installation may entail hidden costs a prospective user should be aware of,
however. Large Ar-lasers (15 to 18 W, all lines) require 460 V,
three-phase line voltage. This is not available in Canada, for instance, where one has 570 V, three-phase. A transformer may
therefore have to be purchased. The water needed for laser cooling

may be too hard, in which case a water softening unit has to be acquired. Large Ar-lasers demand about 75 psi water pressure, which may require purchase of a booster pump.

5.1 WATER COOLING

Water filter cartridges need to be replaced periodically, with frequency depending on the quality of the cooling water. Otherwise, the water flow may become reduced to the point where the laser shuts off. Should the water supply fail altogether while the laser is in operation, the laser tube should be allowed to cool down prior to resumption of water flow to the tube, since addition of cold water to a hot tube can cause damage through thermal shock. Tube cool-down, depending on tube material, may take as much as 1 hr. Some Ar-lasers require continued water flow for about 15 min after laser shutdown. The laser manufacturers' recommendations should be scrupulously followed.

5.2 GAS PRESSURE

Ar-laser tubes operate at gas pressures typically of some 200 mtorr. Prolonged laser operation at pressures much less than this can damage the tube. Operation of Ar-lasers causes the argon gas pressure in the plasma tube to decrease in time. Gas refill systems, which allow the operator to add gas to the tube to bring the pressure back up to the proper operating level, are therefore provided. Tube pressure should be checked only after warm-up of the laser at maximum laser current, and, after each refill, time should be allowed for pressure equilibration. A slight excess in pressure (perhaps 20 mtorr) is not critical. However, high excess in gas pressure can lead to tube failure.

5.3 CLEANING OF OPTICAL SURFACES

While the optical surfaces in Ar-lasers are sealed to avoid contamination, these surfaces (mirrors, Brewster windows) can nonetheless become dirty from dust, smoke, etc., particularly if frequent mirror changes are made. Such mirror changes are involved in change from blue-green to UV lasing or in change from all-lines to single-line operation, and may be a matter of the normal use of the instrument. The use of both UV and blue-green lasing has been described in Chapter 3. Substantial contamination of optical surfaces results in laser power loss, and it may become necessary to clean the laser mirrors and Brewster windows (ends of the laser tube). The coatings of the laser mirrors are very delicate and must not be rubbed, scratched, or touched by fingers. Dust particles on optical surfaces should be blown away with dry air prior to the cleaning procedures about to be described to avoid scratching of the surfaces by these particles. To clean a mirror, a clean piece of lens tissue is placed over the mirror and a drop or two of methanol or acetone is squeezed onto the tissue at the location of the mirror such that the tissue "adheres" to the mirror surface by surface tension. The tissue is then gently drawn across the mirror. If the procedure needs repeating (contamination of a mirror can often be seen by inspection), a new portion of the lens tissue or a new piece of lens paper should be used to avoid respreading earlier removed contaminants. The Brewster windows at the end of the laser tube are less delicate and can be cleaned somewhat more vigorously. A piece of lens paper is folded and held in a hemostat. The tissue is again soaked in methanol or acetone and rubbed with slight pressure across the Brewster window. Care should be taken not to touch the Brewster window with the hemostat. Only one swipe should be made with the tissue. If repetition of the cleaning step is necessary, a new piece of tissue should be used to prevent respreading of earlier removed contaminants. Prisms are cleaned similarly. Brewster windows and mirrors of UV option-equipped lasers require an additional cleaning step with hydrogen peroxide. The laser instruction manual should be consulted for details.

5.4 LASER ALIGNMENT

Ar-lasers are factory aligned and checked out (realigned, if necessary) during initial installation, and should retain their alignment during normal operation. However, if frequent mirror changes are made, the laser might become misaligned, which results in power loss because the laser beam is no longer properly centered in the plasma tube, as shown in Figure 5.1. Both the output mirror (semitransparent) and the rear mirror (totally reflecting, with prism for single-line operation) have screw adjustments which rotate the mirror horizontally and vertically. The following procedure can be used to align the laser (see Figure 5.1). After warm-up, the laser power is measured. The vertical adjustment screw of the output mirror is then turned in one direction by a fraction of a turn. The rear mirror vertical adjustment is then turned to op-

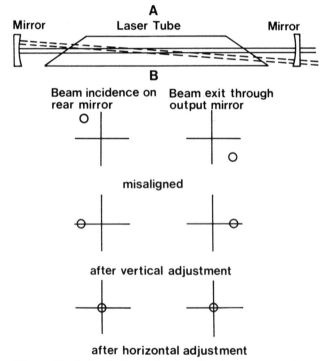

Figure 5.1 (A) misaligned (broken lines) and aligned (solid lines) laser beam. (B) The intersections of lines represent the centers of laser mirrors; the circles denote the location of the laser beam on the mirrors (see text for alignment procedure).

timize the output power. If the optimized power is higher than that prior to the beginning of the adjustments, the output mirror vertical adjustment screw is turned again by a fraction of a turn in the same direction and the output power is again optimized by vertical adjustment of the rear mirror. This procedure is repeated until maximum power is obtained. If the first power optimization step yields lower power than that prior to rotation of the output mirror, then the direction of turning of the vertical adjustment screw of the output mirror is simply reversed. Once vertical adjustment is completed, the same procedure is applied to the horizontal adjustment. Alignment procedures vary somewhat with laser manufacture, and the instruction manual for the instrument on hand should be consulted for details. In some Ar-lasers the laser tube has adjustment screws by which it can be tilted with respect to the laser mirrors. The alignment procedure in this case is similar to that using mirror rotation.

5.5 LASER SAFETY

Ar-laser power supplies and heads are equipped with switches such that the laser shuts off automatically when either the power supply or the laser head is opened while the machine is in operation. These switches have override capability, which allows one to troubleshoot with the laser running. This should be done with caution, since dangerously high voltages exist in the laser head and power supply.

The mandatory use of laser safety goggles during inspection of exhibits under laser light has already been cited in Chapter 3. The New York Laser Safety Code of 1972, for instance, specifies a maximum safe corneal exposure of 10 μW/cm^2 for CW lasers (visible light). This means that laser safety goggles used with large Ar-lasers (of the order of 10 W power at 5145 or 4880 Å) must attenuate the laser light by about a factor 10^7, i.e., laser safety filters have optical density of 7 or larger at the laser wavelengths. For normal (blue-green) Ar-laser operation, laser safety goggles are commercially available (e.g., Fisher 11-409-59A). For UV lasing and, simultaneously, ability to observe luminescence in the

4000 to 5000 Å range (as required for fluorescamine and the phosphorescent powders of Chapter 3), goggles can be made quite simply by replacing the normal orange laser safety filters of a pair of goggles with long-wavelength-pass filters which block wavelengths shorter than about 3800 Å. Corning 3-72 or 3-73 filters, for instance, are adequate. To be on the safe side, each orange filter should be replaced by two UV-blocking long-wavelength-pass filters. Welders' goggles can serve as frames for these filters.

INDEX

105